Tunneling and Underground Transport

FUTURE DEVELOPMENTS IN TECHNOLOGY, ECONOMICS, AND POLICY

Papers prepared for the conference held by the Macro-Engineering Research Group and the Center for Advanced Engineering Study of the Massachusetts Institute of Technology in cooperation with Stone and Webster Engineering Corporation, Arthur D. Little, Inc., and The American Society for Macro-Engineering, April 24 and 25, 1985, at the Copley Plaza Hotel, Boston, Massachusetts

Editor

Frank P. Davidson

Macro-Engineering Research Group
Massachusetts Institute of Technology

ELSEVIER
New York • Amsterdam • London

Thomas Stockebrand about to press the plunger that sent a ping-pong ball (representing a train) at supersonic speed through a 2-inch diameter plastic pipe laid out along the edge of the athletic field of the Massachusetts Institute of Technology. This demonstration was the culminating event of the Conference, **Tunneling and Underground Transport: Future Developments in Technology, Economics, and Policy.** Directly behind Mr. Stockebrand are Dr. Yasuo Sasa, Chief Geologist of Japan's Seikan Tunnel, now "holed through" to connect the main islands of Honshu and Hokkaido; Kenneth E. McIntyre, Brigadier General, U.S. Army (retired), who served as Engineer-Manager of the Tennessee-Tombigbee Canal, the largest project in the history of the Corps of Engineers; and Sir Alan Muir Wood, Senior Advisor to Her Majesty's Government and to private interests involved in the English Channel Tunnel project.

Members of Advisory Committee for Conference on Tunneling and Underground Transport

A.R. Bailey
Former Assistant Deputy
Minister of Supply (Canada)

Robert C. Cowen
Natural Science Editor
Christian Science Monitor

Frank P. Davidson
(Co-Chairman)
Coordinator and Lecturer
Macro-Engineering Research Group
Massachusetts Institute of Technology

Herbert H. Einstein
Professor
Department of Civil Engineering
Massachusetts Institute of Technology

Peter E. Glaser (Co-Chairman)
Vice President
Arthur D. Little, Inc.

Gerald H. Gleason
Vice President
Foxboro Company

John W. Landis (Co-Chairman)
Senior Vice President
Stone and Webster Engineering
Corporation

John Manzo
Corporate Manager
System Development Processes
Digital Equipment Corporation

Kenneth E. McIntyre
Brigadier General
U.S. Army (Ret.)
Manager of Projects
Stone and Webster Engineering
Corporation

J. Lyndon Rosenblad
Assistant Chief Geotechnical Engineer
Stone and Webster Engineering
Corporation

Christine Simonsen
Director
Conference and Seminar Office
Center for Advanced Engineering Study
Massachusetts Institute of Technology

Leonard Unger
Professor of Diplomacy
Fletcher School of Law and Diplomacy
Tufts University

David Gordon Wilson
Professor
Department of Mechanical Engineering
Massachusetts Institute of Technology

Elsevier Science Publishing Co., Inc.
52 Vanderbilt Avenue, New York, New York 10017

Sole distributors outside the United States and Canada:
Elsevier Applied Science Publishers Ltd.
Crown House, Linton Road, Barking Essex IG 11 8JU, England

Library of Congress Cataloging in Publication Data

Tunneling and Underground Transport

Includes index.
1. Tunneling—Congresses. 2. Tunnels—Congresses.
I. Davidson, Frank Paul, 1918- . II. Massachusetts Institute of Technology. Macro-Engineering Research Group. III. Massachusetts Institute of Technology. Center for Advanced Engineering Study. IV. Title: Underground transport.
TA800.T783 1987 624.1′9 87-520
ISBN 0-444-01130-7

Current printing (last digit):
10 9 8 7 6 5 4 3 2 1

Manufactured in the United States of America

Contents

Contributors

William L. Bohannan
Associate Division Director
Taurio Corporation
481 Gold Star Highway
Groton, CT 06340

Edward J. Cording
Professor of Civil Engineering
University of Illinois at
Urbana-Champaign
Urbana, IL 61801

Frank P. Davidson
Macro-Engineering Research Group
Massachusetts Institute of Technology
Room E40-294
Cambridge, MA 02139

Lloyd A. Duscha
Deputy Director
Engineering and Construction
Directorate
Department of the Army
Office of the Chief of Engineers
20 Massachusetts Avenue, N.W.
Washington, D.C. 20314

Herbert H. Einstein
Professor
Department of Civil Engineering
Massachusetts Institute of Technology
Room 1-330
Cambridge, MA 02139

Peter E. Glaser
Vice President
Arthur D. Little, Inc.
15/208 Acorn Park
Cambridge, MA 02140

Kaj Havnoe
Executive Vice President
Technical Department
Christiani and Nielsen A/S
Vester Farimagsgade 41
Copenhagen
DENMARK

Walter J. Hickel
Chairman
Yukon-Pacific Corporation
P.O. Box 101700
Anchorage, AK 99510

Sir Robert G.A. Jackson, KCVO, CMG, OBE, AC
Senior Advisor to the Secretary-General
The United Nations
Room 2520
New York, NY 10017

Martin N. Kelley
President
Kiewit Engineering Company
1400 Kiewit Plaza
Omaha, NE 68131

Thomas R. Kuesel
Chairman
Parsons Brinckerhoff Quade and
Douglas, Inc.
One Penn Plaza
250 West 34th Street
New York, NY 10119

John W. Landis
Senior Vice President and Director
Stone and Webster Engineering Corporation
245 Summer Street
Boston, MA 02107

John W. Leonard
Vice President
Morrison-Knudsen Company, Inc.
Morrison-Knudsen Plaza
P.O. Box 7808
Boise, ID

Wilhelm Linkerhagner
Leiter der Bahnbauzentrale
Hauptverwaltung
Deutsche Bundesbahn
Friedrich-Ebert-Anlage 43-45
Postfach 11 04 23
6000 Frankfurt 11
WEST GERMANY

Cyril C. Means, Jr.
Professor
New York Law School
57 Worth Street
New York, NY 10013

Sir Alan Muir Wood
Senior Partner
Sir William Halcrow and Partners
Vineyard House
44, Brook Green
London, W6 7BY
ENGLAND

Pharo A. Phelps
Manager of Defense Projects
AF Rapid Vertical Boring Egress Project
Bechtel National, Inc.
45 Fremont Street—45/28/B35
P.O. Box 3965
San Francisco, CA 94119

Richard J. Robbins
President
The Robbins Company
P.O. Box C8027
Kent, WA 98031-0427

John C. Rowley
Staff Member
Los Alamos National Laboratory
Los Alamos, NM 87545

Robert Salter
1514 Sorrento Drive
Pacific Palisades, CA 90272

David S. Saxon
Chairman of the Corporation
Massachusetts Institute of Technology
Room 5-205
Cambridge, MA 02139

Harrison Schmitt
Technology and Management Consultant
P.O. Box 8261
Albuquerque, NM 87198

Kenneth Schoeman
Head of Tunnel Section
Division of Dam and Waterway Design
Bureau of Reclamation
Engineering and Research Center
D271 Room 1102
P.O. Box 25007
Lakewood, CO 80228

Carl B. Sciple
Division Engineer
New England Division
U.S. Army Corps of Engineers
424 Trapelo Road
Waltham, MA 02154

Minoru Shimokawachi
Superintendent
Seikan Tunnel Construction Planning Department
Japan Railway Construction Corporation
2-14-2, Nagatachyo, Chiyoda-ku
Tokyo
JAPAN

Charles B. Steward
Assistant Director of Construction
Construction Directorate
MBTA
10 Park Plaza
Boston, MA 02116

Tom Stockebrand
1013 Tramway Lane NE
Albuquerque, NM 87122

Lester C. Thurow
Professor
Sloan School of Management
Massachusetts Institute of Technology
Room E52-455
Cambridge, MA 02139

Paul E. Tsongas
Foley Hoag and Eliot
One Post Office Square
Boston, MA 02109

PART I

SETTING THE STAGE

Historical and Contemporary Experience

Introduction

FRANK P. DAVIDSON
Coordinator of Macro-Engineering Research Group, School of Engineering, Massachusetts Institute of Technology, Cambridge, MA 02139

Nineteen eighty-five was, in several respects, "the year of the tunnel." The great Seikan Tunnel, linking the main Japanese island of Honshu with its northern neighbor, Hokkaido, was "holed through" beneath the treacherous 13-mile-wide Tsugaru Strait. On the other side of the world, Anglo-French intergovernmental deliberations at the highest level led to the designation of twin railway tunnels, with a pilot -- and service -- tunnel in between, as the preferred "fixed link" between the white cliffs of Dover and the equally white cliffs of Calais. Thus, the plan for an underground (and undersea) transport link developed a full generation ago by the Channel Tunnel Study Group came to fruition: on February 12, 1986, a Channel Tunnel Treaty was formally signed, in the Chapter House of Canterbury Cathedral, in the presence of President Mitterand of France and Prime Minister Thatcher of the United Kingdom.

With these and other burgeoning events as backdrop, the conference of April 24 and 25, 1985, entitled "Tunneling and Underground Transport -- Future Developments in Technology, Economics, and Policy," was brilliantly timed. The sponsoring institutions had determined that this would be more than a run-of-the-mill conference: preoccupation with outer space and computer technology had left the seemingly pedestrian field of tunneling wihout comparable championship and visibility. A public somewhat inclined to regard both technology and politics as branches of the entertainment industry had failed to grasp the inevitable interplay between "high tech" and "low tech," between aerospace and surface transport on the one hand, and the corresponding role of underground space and subterranean transport on the other. There were vital issues to be addressed, some of them barely formulated. What seemed necessary, in the view of the organizing committee, was a serious forum where seasoned professional engineers could explain to leading specialists in other fields what had been accomplished and what is expected in the domain of technology, and where respected economists, government officials, and opinion leaders could present their own views on the present and prospective status of an engineering activity which goes far back beyond the dawn of recorded history.

This multilateral character of the conference was prefigured by the composition of the sponsoring group itself and its designated organizing committee. The Massachusetts Institute of Technology (M.I.T.) initiated the program through the Center for Advanced Engineering Study and the Macro-Engineering Research Group, both in the School of Engineering. Stone and Webster Engineering Corporation, builder and long part-owner of Boston's subway system, joined the team and provided detailed intellectual and logistical support. Arthur D. Little, Inc., an interdisciplinary consulting firm founded more than a century ago by two graduates of M.I.T. and once owned outright by M.I.T., also came loyally aboard. A final and valued sponsor was the American Society for Macro-Engineering with headquarters at Polytechnic University in New York.

The organizing committee exemplified the twin principles of expertise and diversity. A.R. Bailey, the renowned Canadian civil servant, marshalled the considerable competence of Canadian firms which have pioneered new and now proven methods of excavation. Robert C. Cowen, Science Editor of The Christian Science Monitor, attended to the vital area of press relations. Professor Herbert H. Einstein of M.I.T.'s

Tunneling and Underground Transport: Future Developments in Technology, Economics, and Policy
F.P. Davidson, Editor

Department of Civil Engineering, brought to bear his wide contacts as a member of the United States National Committee on Tunneling Technology of the National Academy of Sciences. Gerald H. Gleason, Vice President of the Foxboro Company, and John Manzo, the Corporate Manager of System Development Processes at Digital Equipment Corporation, contributed their "high tech" prowess to discussions on what had been downgraded (implicitly) as a "low tech" subject! We were fortunate also to obtain the participation of Kenneth E. McIntyre of Stone and Webster Engineering Corporation. As a Brigadier General in the United States Army Corps of Engineers, he had supervised construction of the largest project in the history of the Corps: the Tennessee-Tombigbee Canal. J. Lyndon Rosenblad, Assistant Chief Geotechnical Engineer at Stone and Webster Engineering Corporation, contributed his unsurpassed knowledge of the professional leadership of the tunneling industry, and was a key participant in the actual process of planning the conference. It was a stroke of luck, also, that we were able to wedge our activity into the busy schedule of Dr. David Gordon Wilson, a professor of mechanical engineering at M.I.T., who invented the P.A.T. system (Palleted Automated Transit) which was successfully exhibited in the form of a model contrived by his adroit student, Brian Barth, and placed in the assembly hall of the conference at the Copley Plaza Hotel.

Now that tunneling has taken on international dimensions, it seemed appropriate to incorporate the advice and assistance of a senior diplomat. Professor Leonard Unger of the Fletcher School of Law and Diplomacy brought a useful planet-wide perspective to bear on our deliberations. A former United States Ambassador to Taiwan and Thailand, he reminded us that the intercontinental links foretold in the short stories of Jules Verne may, in the next century, become objects of close scrutiny from both the diplomatic and the economic viewpoints. Christine Simonsen, the experienced Director of the Conference and Seminar Office at the Center for Advanced Engineering Study, M.I.T., completed the roster of committee members. She was very ably assisted by Barbara Dullea, Mary Powers, Allison Cocuzzo, Cecil Feaver, and Michael Quinlan. In the "run-up" to publication of these proceedings, special mention must be made of the superb talent and devotion of Mary Powers, who assembled the "raw materials" from scattered and not-always-so-available contributors!

For the complex word processing and production of the manuscript in "copy-ready" form, much appreciation is surely due to Cherie Potts, Word Processing Supervisor at M.I.T.'s Sloan School of Management. She was able to take time out from an unbelievably demanding schedule to see this book through its myriad prepublication details. An Assistant Editor with experience in previous macro-engineering volumes, she was the indispensable link between an itinerant general editor and the irreducible exigencies of a (necessarily sedentary) press. Of course, the whole affair owes much to both the practical wisdom and the human and professional sensitivity of George V. Novotny, Vice President of Applied and Information Sciences at Elsevier Science Publishing Co., Inc. George is an afficionado of macro-engineering. He was quick to perceive the public importance of what might have appeared to undiscriminating observers as just one more academic-type symposium. Barbara Schwagerl, the Elsevier Desk Editor, was meticulous, constructive, and a veritable "Rock of Gibralter" in overseeing the final details of collation and presentation.

I cannot say enough in praise of my two co-chairmen of the conference: John W. Landis and Peter Glaser. Mr. Landis served in a dual capacity: as President of the American Society for Macro-Engineering and as Senior Vice President of Stone and Webster Engineering Corporation. Dr. Glaser, who was able to call on the impressive knowledge of Albert J. Kelley in construction industry matters and on the fantastic outreach of

Ms. Alma Triner in public relations questions, himself brought an added "level" to our discussions: as inventor of the Solar Power Satellite concept and the holder of its principal patent, Dr. Glaser served as a living bridge between the earthbound "moles" at the meeting and those forerunners of the next millenium who have either visited the moon (Harrison Schmitt), or expect to! We were fortunate to have Dr. David Saxon, Chairman of the M.I.T. Corporation, to introduce Dr. Schmitt to the dinner audience on the first day of the conference. And just as the geologist-astronaut led us to ponder the application of our local civil engineering capabilities in the wider arena of outer space, so Sir Robert G.A. Jackson, the defender of Malta in World War II, reminded us of the role of tunneling in warfare.

The conferees were, understandably, oriented toward an analysis of recent experience, hopefully as a guide to impending projects and events. It was well, therefore, that the first luncheon speaker, Professor Cyril C. Means, Jr., put the whole subject of tunneling in its proper historical context. That an expert on constitutional law should have acquired such detailed knowledge of ancient tunnels may appear puzzling. But this research had been mandated, in fact, by the situation of the channel tunnel project, in whose revival after World War II, Professor Means played a conspicuous role. Sir Pierson Dixon, former British Ambassador in Paris, had written a novel in which one of the heroes was Julius Ceasar's chief engineer, who happened to be a tunnel specialist. It was with a view to reinforcing Sir Pierson's historical orientation on channel tunnel affairs that Professor Means, more than twenty years ago, carried out his inquiries into the tunnel feats of ancient Egypt, Persia, Greece, and Rome.

The conference papers imply, but they do not directly represent, the cross-fertilization of ideas and experiences that took place in the coffee breaks, the lunches and dinners, and the other events associated with the formal sessions. Credit is due Ms. Suzanne Fairclough, Director of the M.I.T. Macro-Engineering Film Collection, for organizing such an enticing array of documentary films for times when it seemed as well to look as to listen. And, of course, the final event of the conference turned out to be somewhat spectacular: the first known public demonstration of supersonic flight through an evacuated tube. (The word "supersonic" is used to indicate velocity: sonic booms do not constitute a problem in a vacuum!) It has been said that Thomas Stockebrand's demonstration, on the M.I.T. Athletic Field in the late afternoon of April 25, 1985, was perhaps the most portentous event in the history of transportation since the flight of the Wright brothers' airplane at Kitty Hawk in 1903. We have left, unvarnished, Mr. Stockebrand's account of the exploit, for which he received an Isambard Kingdom Brunel Award. (The award was presented at a ceremony in the Harvard Faculty Club on December 10, 1985, by Dr. Stuart Smith, Chairman of the Science Council of Canada, who flew from Ottawa to Cambridge for the occasion.)

Without attempting to prejudge the question or to pin a "time frame" on future decisions and developments, tunnelers must now, if only subliminally, take into account the possibility that investors will eventually become interested in very long-distance tunnels. Three factors suggest this possibility: first, the development of transport technologies suitable for tunnels and tubes and ultimately capable of velocities superior to those of present-day aircraft; second, the increasing congestion and pollution of surface and air space; and third, the preoccupation with safe installations in the event of war. Clearly, the high cost of tunneling will prevent any very rapid realization of these new and tantalizing speculations. But the ten-year study program launched in Japan and South Korea to find solutions to the technical, financial, and administrative problems of a tunnel-and-causeway complex across the Sea of Japan to the mainland of Asia should cause us to beware of a too-confident scepticism.

The near-term future may offer novelty enough: as subways proliferate, there will be increasing interest in "slotting in" high-tech capabilities so that underground tunnels can carry motor traffic, containers and, where routes are appropriate, garbage. It has even been suggested that countries of continental dimensions such as the United States may devise ways to finance cross-country sportsways--for hiking, biking, horseback riding, and skating--by placing utilities' tunnels underneath a landscaped and protected right-of-way. Thus, conservationists, fitness enthusiasts, and regional planners would be challenged to cooperate with industry to provide nonpolluting, fail-safe, aesthetically acceptable routes for water mains, oil and gas pipelines as well as slurry pipelines, electric power lines, fiber-optics lines for telecommunications, and conceivably pneumatic tubes for the ultra-rapid transport of mail and parcels. It is truly a "large conception," very much in the tradition of Benton Mackaye's original proposal, in 1922, for an Appalachian Trail.

The "Tunneling and Underground Transport" conference encouraged lively exchanges of views among bankers, lawyers, journalists, public officials, business leaders, and engineers. It was tacitly accepted that macro-engineering projects are now so costly and of such pervasive importance to society that interprofessional meetings and study groups are warranted. Above all, thanks are due to the attendees, drawn from many professions and indeed from many countries, who made this gathering a most memorable and encouraging event.

Dr. Glaser and Mr. Landis (my fellow co-chairmen) and all the members of the organizing committee labored hard and diligently to make the meeting significant to the wider public; all of us must acknowledge a very deep obligation for their devotion and perspicacity. The presentations in this volume represent the essence of what was said at the meeting: _res ipsa loquitur_. If the reader of the chapters that follow can catch a modicum of the excitement of those who attended the conference, the editor and his associates will feel amply compensated for their efforts. A civilization that reaches for wider contacts in the universe will not fail to make fuller use of that oft-ignored asset, underground space on our own planet.

Ancient Tunnels

Professor Cyril C. Means, Jr.
New York Law School
57 Worth Street
New York, NY 10013

In November 1963 President de Gaulle dispatched an emissary to Peking to sound out Mao Tse-Tung and Chou En-lai on the possibility of normalizing diplomatic relations between France and China, a step which both countries agreed on two months later. The man he chose for this delicate mission was Edgar Faure, a former Prime Minister and one of those political intellectuals whom France produces more examples of than other countries. Faure's mission--like Henry Kissinger's eight years later--was intended to be strictly secret, but, of course, the Paris press got wind of it, so a throng of reporters greeted Senator Faure when he landed at Orly on November 20, 1963. "What did you learn in Peking?" was the single query thrown at him by a score of voices.

Faure paused, then answered. "Gentlemen: In China I learned that the Chinese preceded us Europeans by centuries in some of our greatest inventions. To name but three examples: gunpowder, the astrolabe, and movable type. But the Chinese never exploded gunpowder to wage war, never employed the astrolabe to discover America, and never used movable type to print . . . newspapers!" Stunned, the reporters stood in abashed silence while Faure proceeded, without another word, to his waiting limousine.

Good as Faure's bon mot was, for the purpose for which he used it, it contains a deeper meaning, which bears a bit of pondering.

The Chinese are not alone among ancient peoples in having independently discovered great inventions, which they put to an entirely different use. It is often said that the Aztecs never invented the wheel, and first learned of it when they saw the Spaniards' carriages and carts. It is true that adult Aztecs had no wheeled carriages or carts. But Aztec children had toy carts with tiny wheels!

Apparently, it never occurred to the Aztec parents who gave these gifts that wheels were suitable for adult purposes as well as for children's toys. Yet it should be remembered that the Aztecs possessed astronomical skills comparable to those of the ancient Egyptians.

If the Chinese used gunpowder only for fireworks, the astrolabe only for astronomical calculations, and movable type only for producing very limited editions of very brief texts, and if the Aztecs used wheels only as toys, the reason must be that the ends to which they devoted these inventions were those that interested them, rather than those that interested us. Religious and cultural differences undoubtedly played a major role in shaping what those interests were.

We then may ask: How do tunnels fit into this picture? Have they been used for startlingly different purposes by different national and cultural groups?

In order to sketch an answer to this question, let us recount the stories of six famous tunnels of antiquity: one built by Babylonians, a second by Hebrews, a third by Greeks, and three more by Romans.

The earliest of these ancient tunnels was in Babylon, under the Euphrates. It provided a passageway for pedestrians between the royal

Tunneling and Underground Transport: Future Developments in Technology, Economics, and Policy
F.P. Davidson, Editor

palace and the temple, which were on opposite banks of the river. Lined with brick, it was 3000 feet long. Construction was accomplished by diverting the river during the dry season, between 2180 and 2160 B.C., more than 41 centuries ago.

The second oldest of our six tunnels was built in and near Jerusalem, about 704 B.C., by Hezekiah, the thirteenth king of Judah. His purpose was to secure a supply of water to the Holy City during an expected siege by the Assyrians, which actually took place three years later, after Sennacherib, their new king, had defeated the Babylonians and the Egyptians.

The only natural spring was at Gihon, emerging from a cave in the Ophel Hill outside the city's east wall. The tunnel was angled to bring water, by gravity, to the Pool of Siloam in the southwest corner of the city, well within the walls. Had Hezekiah's engineers succeeded in following a straight-line course, the tunnel would have been only 1100 feet long. Many obstacles they encountered, however, forced them to change course a number of times, increasing the final length to 1750 feet. Without facilities for measuring and recording these deviations, the two teams got lost several times, and had to drive vertical shafts to the surface to ascertain where they were in relation to each other. The cave at the source was sealed up to prevent enemy access to, or tampering with, the water supply. The tunnel has remained intact to this day.

Sennacherib boasted that he had imprisoned Hezekiah in Jerusalem "like a bird in a cage," and so he had, but with the tunnel in place, neither Hezekiah nor his subjects would die of thirst; so eventually the Assyrians departed and the siege was lifted.

In 1880, an inscription was discovered in the rock wall of the tunnel, describing how it was excavated, fleshing out the sparse biblical references to it in 2 Kings 20:20 and 2 Chronicles 32:30. This inscription is the oldest known example of cursive Hebraic written in the Phoenician alphabet. The language is perfect classical Hebrew prose typical of Hezekiah's time.

Our third tunnel of antiquity was built on the island of Samos, in the Sixth Century B.C., to carry water from a well through a mountain to the city of Samos. As in all ancient water tunnels, conduction was effected by gravity flow. This tunnel was six feet wide and six feet high, and 3400 feet long. Herodotus identified the engineer as "Eupalinus, son of Naustrophus, a Megarian."

While Herodotus praised the completed tunnel as among the "greatest works to be seen in any Greek land," serious errors were committed in constructing it. The two teams managed to miss each other by sixteen feet, an angling error of 0.5%. The surveying devices available to Eupalinus should have assured a much closer encounter.

During the Middle Ages, the Samos tunnel fell into disuse, but it was rediscovered in A.D. 1882, and is once again in use.

The fourth and fifth of our tunnels are actually a pair of parallel highway tunnels pierced through the massive Ridge of Posilippo, which separates Naples from the spit of land on which stood the Imperial Villa Pausilypon, and also the small town of Puteoli, now called Pozzuoli. Before these tunnels were built, the only way of travelling between Naples and either the Villa or Puteoli was by boat.

In 36 B.C., Marcus Agrippa, co-consul of the Emperor Augustus, caused the larger of these tunnels to be built by the engineer Marcus Cocceius Nerva. This tunnel, today called the Grotta di Sejano, was intended for the service of the Imperial Villa. The width varies from 13 to 21 feet; the height from 14 to 28 feet. It is 2526 feet long. In the Fifth Century, A.D., this Grotta di Sejano ceased to be used, and its portals became blocked with earth. It was rediscovered and reopened in 1840-1841.

The less commodious of the two Naples-Pozzuoli tunnels was described by Seneca. It lacked the shafts of light and air of its imperial counterpart, being intended for use by ordinary people. Today it is called the Grotta di Posilippo. In width it varies from 8 to 10.3 feet; in height (at mid-passage) from 8.6 to 17 feet. It is 2322 feet long.

Constructed somewhat later than the one leading to the Imperial Villa, the Grotta di Posilippo has been in continuous use since it was opened. At an age of 19 centuries, it is the oldest continuously used traffic tunnel in the world.

The sixth and most recent of our ancient tunnels was dug in the decade 41-51 A.D., for the purpose of draining Lake Fucino, north of Rome. Half a million cubic feet of rock were removed to create an emissary that reclaimed 50,000 acres for the imperial estates. Under the orders of the Emperor Claudius, the engineer Narcissus created this longest tunnel of ancient times--it was 3.5 miles long. It measured 9 feet in width and 19 feet in height.

Stopped up through disuse during the Middle Ages, it was reopened when, between 1854 and 1875, Prince Torlonia drained the lake, whose former bed is now dry land.

Like nearly all ancient tunnels, the Lake Fucino Emissary was dug by slaves. Suetonius puts their number at 30,000, and states that they worked on the project eleven years without interruption. Pliny the Elder declared that it "cost a sum beyond all calculation," and was accomplished under conditions so grueling that "no human language can possibly describe" them.

After the project was thought to have been completed, Suetonius describes two episodes that undoubtedly are far more amusing to us than to those who participated in them. Claudius arranged for an opening celebration worthy of so major an engineering feat. 19,000 prisoners were compelled to participate in a gigantic sham naval battle on the lake, just before pulling the plug and draining it. The Praetorian guard ringed the shore to prevent the participants from escaping. The battle went off according to schedule, whereupon the Emperor signalled that the sluices to the tunnel should be opened. Then sore malfunction occurred: not a drop of water entered the tunnel. The Emperor and his party rode back to Rome in fury over the failure.

Narcissus was fortunate: he was given another chance. After another year spent deepening and improving the grade of the emissary, another opening party was arranged. At the downstream end of the tunnel tables were set for a banquet. After a gladiatorial contest, the Emperor again signalled that the sluices should be opened at the lake. This time everything worked--too well, in fact. The water came gushing out of the downstream portal with such force that the tailrace canal became flooded. The banquet tables were swept away. The Empress Agrippina screamed at Narcissus, accusing him of graft and much else. Narcissus retorted in kind, casting heavy aspersions on the Empress's lifestyle, which was, indeed, "no better than it ought to be."

As we look back upon our six ancient tunnels, do they give us any evidence that different nations or cultural groups used tunnels for widely divergent purposes, as, for instance, the Chinese and the Europeans did the three inventions mentioned by Senator Faure, or as Aztecs and all other peoples did the wheel? The answer must, I submit, be No.

Tunnels have been used by all nations and cultural groups, ancient and modern, for the same handful of purposes: mining, water supply, drainage, pedestrian and vehicular traffic. But each group has used tunnels for all these purposes. No spectacular variations in use attributable to cultural or national difference have appeared.

The tunnel, therefore, is a sign, not of the diversity, but of the unity, of mankind.

Current Applications of Tunneling

THOMAS R. KUESEL
Chairman of the Board
Parsons Brinckerhoff Quade & Douglas, Inc.,
One Penn Plaza
250 West 34th Street
New York, NY 10119

Much of the focus of this meeting will be directed toward transportation tunnels, but before we get down to some of the specific examples, I thought I would start the conference with a brief survey of the great variety of projects to which tunneling has been and is being applied. Perhaps this will stimulate some ideas on how we might expand the scope still further in the future.

Let us start close to home. Figure 1 shows the Harvard Square station, across the river in Cambridge, part of the Boston MBTA Red Line system which will be discussed in more detail by Mr. Steward. This is what is called "the shoehorn method." The existing line runs down Massachusetts Avenue, and the extension curves out through Harvard Square to the north with very tight quarters. It was necessary to put the new wall construction within a couple of feet of Lehman Hall, part of the Harvard University campus, and there was quite a cliff in front of Lehman Hall. Underneath, it looked like Figure 2 during construction, where we were extending the new line off toward the left foreground while maintaining service on the existing platform in the background.

I should add that we undertook this project in association with Skidmore, Owings and Merrill (SOM). SOM's partner-in-charge was Peter Hopkinson, who went to Princeton. I went to Yale. So for Peter and me, this was the crown of our professional careers, to really demolish Harvard Square.

It is not always that tight, however. Figure 3 shows the Peachtree Center station on the MARTA transit system which was mined out of very solid rock beneath the central business district of Atlanta to make a cavern station with exposed rock walls. The architect was so impressed with the rock that we decided to use it as an architectural material as well as a structural material. It turned out to be both economical and visually striking. (This was inspired, incidentally, by the great Tumut powerhouses of the Snowy Mountains hydroelectric project in Australia, which I visited a year or two before we undertook this design.) Figure 4 is a view during construction, in which all of the construction access was off street (left foreground and right background), and so it was possible to build this enormous subway station right underneath the business district without any disruption of existing traffic and access to the adjacent buildings. It helped, of course, to have good rock underneath.

The largest transportation tunnel in the world under present active status is the Seikan Tunnel in Japan (Fig. 5). Mr. Shimokawachi will give us a detailed discussion of that a bit later. I will make just a brief remark. The water is nearly 500 feet deep and the tunnel another 300 feet below that. (The depth of the tunnel beneath the surface is somewhat greater than the height of the John Hancock Tower.) I will only

Tunneling and Underground Transport: Future Developments in Technology, Economics, and Policy
F.P. Davidson, Editor

Figure 1. Harvard Square Station, Cambridge.

Figure 2. Harvard Square station construction.

Figure 3. Peachtree Center station, Atlanta.

Figure 4. Peachtree Center station construction access.

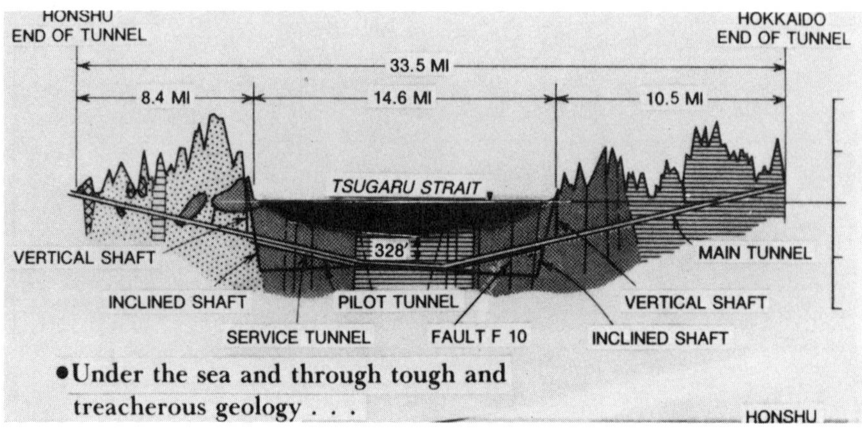

Figure 5. Seikan Tunnel, Japan.

say that my Japanese colleagues have more nerve than I do, because when they were digging down some 500 feet below the ocean, they encountered hot water, which is bad news for tunnelers. So, they calmly drilled a hole up into the ocean to get cold water to mix with the hot water, to cool the tunnel, and then pumped the two of them out.

The largest transportation tunnel in North America underway at present is the Rogers Pass rail tunnel for the Canadian Pacific Railway (Fig. 6). Construction started in earnest just this past year. It will be a four-year project. It is nine miles long through the Selkirk Mountains at the heart of the Canadian Rockies. This is the avalanche capital of North America where the snow cover slides in Rogers Pass. The tunnel runs underneath Mt. MacDonald, beneath the pass, and then underneath Cheops Mountain. In operation, it is two tunnels in tandem. In order to increase the throughput capacity for trains, we put a vertical shaft 1300 feet deep in the middle of the pass, which provides ventilation supply and exhaust; it's a split shaft. In conjunction with that, in the center of the tunnel is a door. The train comes in from the east side and passes through the door. We close it down behind the train, and then blow all the diesel exhaust and smoke out the back half of the tunnel while drawing fresh air in from the front to cool the engine as it goes through the second half. That way we keep the train going without overheating, and we purge the tunnel behind it so that the following train can enter more rapidly. Figure 7 shows the tunnel boring machine being brought into the east portal.

The largest American tunnel project in terms of cost and scope is Baltimore's Fort McHenry Tunnel (Fig. 8), which will be described to you by Mr. Kelley. This had a construction cost of $750 million, half of which was required to pass underneath Baltimore Harbor, and the other half was required to stay down out of sight of Fort McHenry and preserve

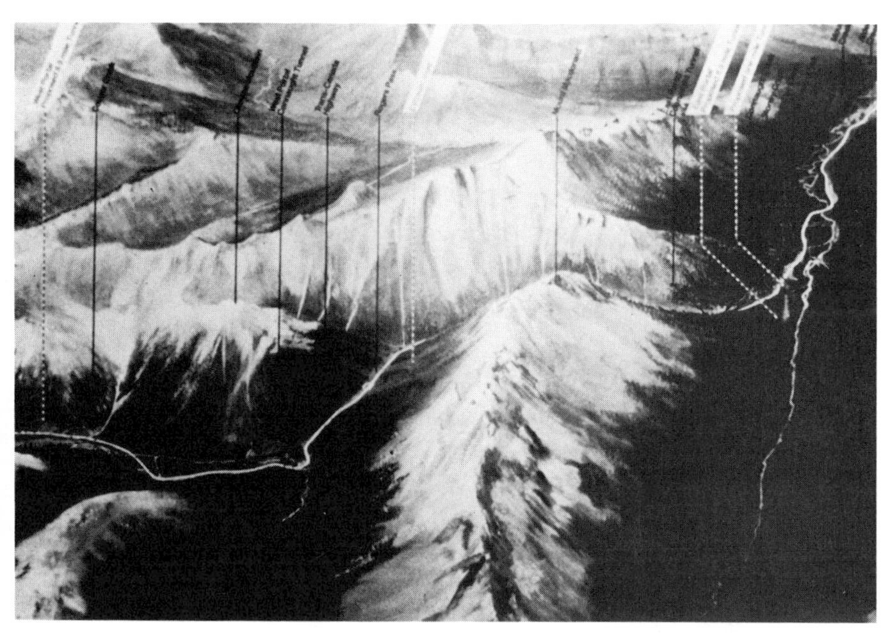

Figure 6. Rogers Pass Tunnel, British Columbia. (Reproduced with permission C.P. Rail.)

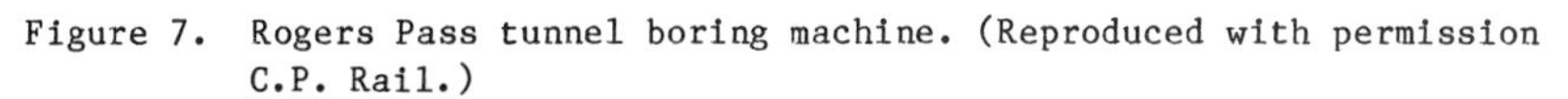

Figure 7. Rogers Pass tunnel boring machine. (Reproduced with permission C.P. Rail.)

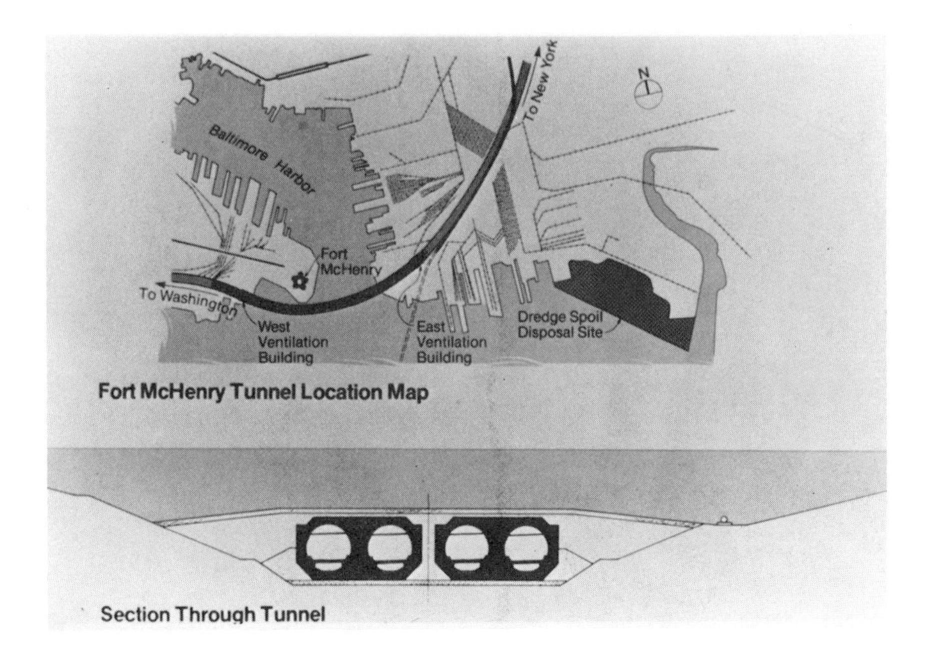

Figure 8. Ft. McHenry Tunnel, Baltimore.

domestic tranquility at the historic monument. A considerable investment, about $300 million, was required to extend the tunnel, but this was the price for gaining permits, approvals, and public acceptance after some 15 years of arguing about the project. This immersed tube type tunnel consists of twin sections, each carrying double bore tunnels, and providing a total of eight lanes. The whole thing is 180 feet wide, and the immersed tube section is about 5,000 feet long. There are cut and cover tunnel sections beyond that extend the total length to 7,200 feet. Figure 9 shows the fabrication of a typical binocular steel shell section at a shipyard up on the Susquehanna River. Figure 10 shows the most exciting part of the construction, which is getting the tubes through the railroad bridge without knocking the railroad out. Figure 11 shows the tube in the lay barge, a catamaran arrangement from which it is suspended while ballast is put into a central pocket to overcome buoyancy and sink (or "immerse") it into place on the prepared foundation in the dredged trench below.

Figures 12 and 13 show the Hampton Roads project at Norfolk, Virginia where the bay was 30 miles wide. The problem of ventilating the tunnel with heavy traffic over that length was very difficult, and the cost was enormous. So back in 1957 a departure was realized, which was to build two islands in shallow water on either side of the channel and to shorten the tunnel to one and a half miles, with two miles of bridges connecting to shore on either side. This is a means of bringing long estuary crossings within financial, as well as engineering and construction, practicality. A number of examples have been promulgated since then.

Turning now to water tunnels, and Mr. Schoeman will have further examples for us later, Figure 14 is the famous New York City Water Tunnel #3, which has been under construction for at least 15 years and under litigation for almost as long. But a section was holed through on

Figure 9. Ft. McHenry Tunnel--shipyard fabrication.

Figure 10. Ft. McHenry Tunnel--towing tube section.

Figure 11. Ft. McHenry Tunnel--tube in bay barge.

Figure 12. Hampton Roads Tunnel approach.

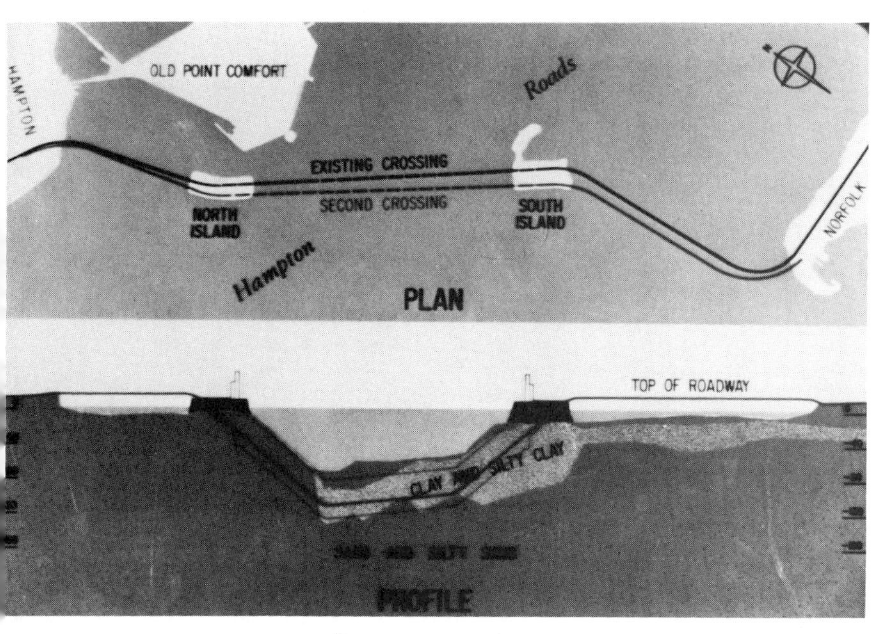

Figure 13. Hampton Roads Bridge--tunnel layout.

Figure 14. New York City Water Tunnel #3.

February 11, 1985 at 3:08 p.m., a historic occasion. I show it because it is a good example of a drill and blast rock tunnel with a concrete lining, which is the traditional way of constructing hard rock tunnels.

In Chicago, the Tunnel And Reservoir Plan (TARP) project is a massive storm drainage project. The basic idea is to get the water out of all the living rooms and playrooms of Chicago where it is presently stored after a good rain. These tunnels were constructed with a hard rock tunnel boring machine (TBM), or a whole series of them, and Figure 15 shows the business end of the cutter head coming through the rock wall. Figure 16 shows a raw excavated tunnel in limestone drilled out by that machine. It looks as though it is a smooth concrete lining. In fact, I took an experienced tunnel engineer into a similar tunnel in Washington, and he swore it was concrete until we walked far enough to pick up some rock joints. So this shows the sort of thing that can be done with modern technology.

Water projects can get very complex. Hydroelectric power projects can generate very large and complicated configurations of tunnels. Figure 17 shows one in California, the Helms Pumped Storage Project, which is reversible. During periods of heavy electric demand, they run it forward, draining water out of the upper lake through the powerhouse down below. During the off-peak hours at night when there is excess power on the line, it is not economical to shut the base-load plants down, and they turn the turbine around and run it backwards to pump the water back up to the lake, so they can use it again the next morning. Figure 18 shows the Helms powerhouse under construction. For scale, there are two men standing under the concrete bucket at lower left. It is 143 feet high, 88 feet wide and 300 feet long. It is a very impressive underground cavern, and this is by no means an unusually large cavern in the underground hydro business. It shows the size of openings that can be created. You can make tunnels out of precast pipe. Figure 19 shows a water project in

Figure 15. TARP boring machine, Chicago. (Reproduced with permission of Harza Engineering Company.)

Figure 16. TBM Tunnel, TARP Project. (Reproduced with permission of Harza Engineering Company.)

Figure 17. Helms Pumped Storage Project, California. (Reproduced with permission of Pacific Gas & Electric Co.)

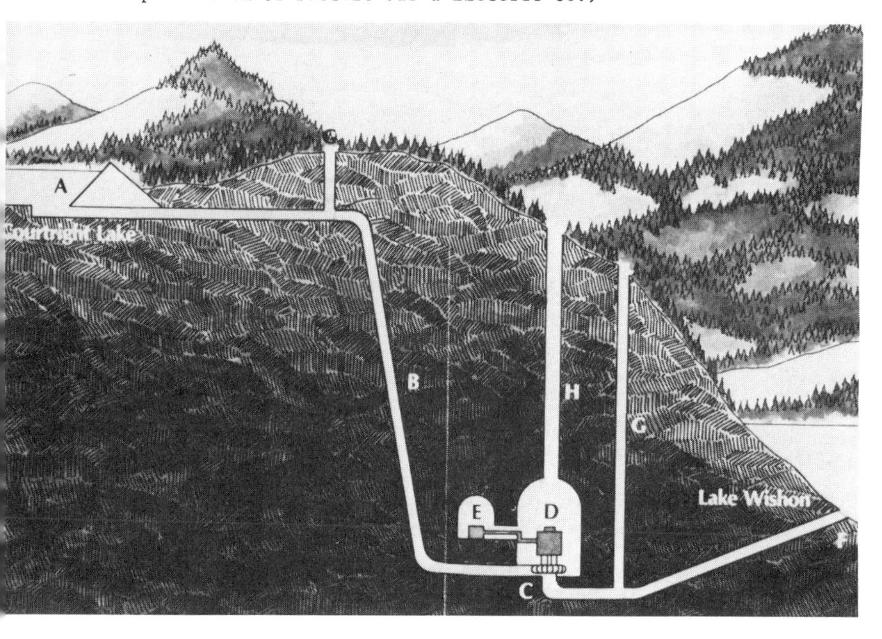

Figure 18. Helms powerhouse. (Reproduced with permission of Pacific Gas & Electric Co.)

Figure 19. Precast pipe tunnel, California.

California, and for scale there is a man walking on top of the pipe. This one happens to be for water; you could easily run a subway train or other forms of transportation through it. Precast pipe sections can be made 20 feet in diameter--this is pretty close to that.

Pipe tunnels are not necessarily confined to land. Figure 20 shows the San Francisco Southwest Ocean Outfall Project (SWOOP) which collects all the waste water, or will when the project is hooked up from the City of San Francisco, and transports it four miles out into the Pacific Ocean to a point where currents will disperse it further out to sea. The photo shows the construction trestle through the surf zone, which is unusually violent. The fetch at this location is 8,000 miles from the Phillipines. The waves can be impressive. Figure 21 shows a 12-foot-diameter pipe section being lowered to make a tunnel beneath the ocean.

We have a whole new class of tunnels spawned by the physics research community. Figure 22 shows a general view of SLAC, the Stanford Linear Accelerator Center, in California, where perhaps 15 years ago they built a two-mile tunnel (coming in from the left), where they accelerate electron particles to the speed of light and send them crashing into targets in these buildings, and measure all kinds of marvelous things about the intrinsic nature of matter. One of the curiosities of this set-up is that at the ends of these buildings you need something to absorb the energy. They settled on a remarkable form of shielding. The Navy had a whole bunch of 16-inch naval guns left over from scrap battleships. They simply collected those and stacked them like cordwood across the ends of the target buildings. So that is swords into plowshares--they found a use for 16-inch naval guns. At any rate, a few years ago the physicists got tired of dealing in a straight line and decided they would do better in a circle, and so they devised a scheme to put magnets on the end of this two-mile run, and run the electrons around a circular tunnel about half a mile across. Then they generated positrons, which are the opposite of electrons, and they run those around at the speed of light in the opposite direction. And then they put some more magnets on them and draw the two beams together and they meet at something like twice the speed of light. My physics did not go that far, but they have done remarkable things with that out in California. This of course merely generates the need, the urgency for yet greater experiments of higher power, which means larger machines and larger tunnels, and they are now planning the SSC, the Superconducting Super Collider, which will have a diameter of 20 miles or more, a tunnel 60 miles long to race these particles around and get the energy to conduct high-energy physics experiments. I have not really understood where they are going to get $4 billion to get this device together, but that certainly qualifies as a macro-project.

Figure 23 shows the machine used to bore the SLAC tunnels, variously called an Alpine Miner, a roadheader, or a pivoted boom excavator. I call it a rotating horse chestnut. It is very effective in moderately soft rocks, and it produces a tunnel (Fig. 24) of any arbitrary configuration. They do not have to be circular. This tunnel is lined with shotcrete. It had a limited service life, and this was sufficient for the purposes of the use of it.

Another class of tunnels is defense facilities or hardened underground facilities. Mr. Duscha will be talking more about this. Figure 25 shows NORAD, the North American Air Defense Command Center, which is now getting

Figure 20. SWOOP, California.

Figure 21. Placing pipe section in the ocean, SWOOP.

Figure 22. Stanford Linear Accelerator Center, California.

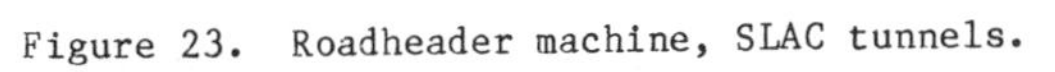

Figure 23. Roadheader machine, SLAC tunnels.

Figure 24. Shotcrete lined tunnel, SLAC.

Figure 25. NORAD tunnels, Colorado Springs.

to be ancient in this business. This goes back 20 years. Under the Cheyenne Mountain in Colorado Springs is a complex of underground chambers, all covered over with rock bolts and wire mesh to contain fragments.

The Strategic Petroleum Reserve in Louisiana and Texas was conceived about ten years ago to give us some margin against eruptions in the Mideast oil market. Here enormous caverns have been constructed in the salt beds of the Gulf Basin by a process of leaching, which is similar to the old Frasch process of mining sulphur. The concept (Fig. 26) is that hot water is pumped down into the salt beds to melt the salt. As water is pumped in at the bottom, the brine is extracted out at the top and gradually enlarges the section. It is kept filled with oil to preserve some internal pressure so the roof does not collapse. Gradually this is extended by putting water in either at the top or the bottom. Adding water at the top widens the cavern out, and draws the brine out at the bottom. Eventually the result is a very large cavity filled with oil, as shown in Figure 27. This is an existing cavern in Louisiana, approximately twice the height of the Empire State Building, and full of oil.

Our Swedish friends have made many advances in tunneling. The Scandanavian climate favors underground protection against winter weather, and they are blessed with very good rock. There are a whole series of oil storage projects in Sweden (Fig. 28) in which they transfer oil directly from the tanker down through a pipe system into a series of underground galleries. Figure 29 shows one of them, not necessarily the largest, but certainly the best lighted. These are 100 feet or more in height, 50 feet wide, and they have within them a complex series of piping for controlling the water and the oil. This is devised so that the ground water maintains an inward pressure and keeps the oil from leaking out into the surrounding environment.

We turn now to habitation, and this is an ancient use of tunnels. Figure 30 shows some that really are not so ancient--they are currently occupied in Tunisia. Professor Means will give us more examples of ancient tunnels, but they are not all confined to the "Old World."

There is a place in Minneapolis where they have taken a great interest in underground work. Figure 31 shows the entrance to Williamson Hall, which is an underground bookstore and university office building set into a triangular excavated area in the middle of the University of Minnesota campus. Figure 32 is taken inside. There are three stories underground here with daylight streaming in. This is a very energy efficient, economical operation as well as a striking environment.

Do you remember the old Sod Busters in Nebraska who built their homes out of dirt? The tradition seems to linger. Figure 33 also relates to Minneapolis, on a main arterial street that is very noisy and heavily trafficked, and not a desirable housing area. But they are covered in one block with a big earth berm. Would you believe, town houses? And this has now become a desirable neighborhood. They are very quiet, no noise from the freeway. They certainly have low heating bills, and they are rented with no trouble at all.

Then you have things like Figure 34. This is in Florida, and is called a "dune house." An architect lives there. I am not sure that I would, but there is no accounting for tastes.

Some of the most remarkable usage of underground space is in Kansas City, where there is a large series of limestone mines that have been economically played out. They have been adopted by the Great Midwest

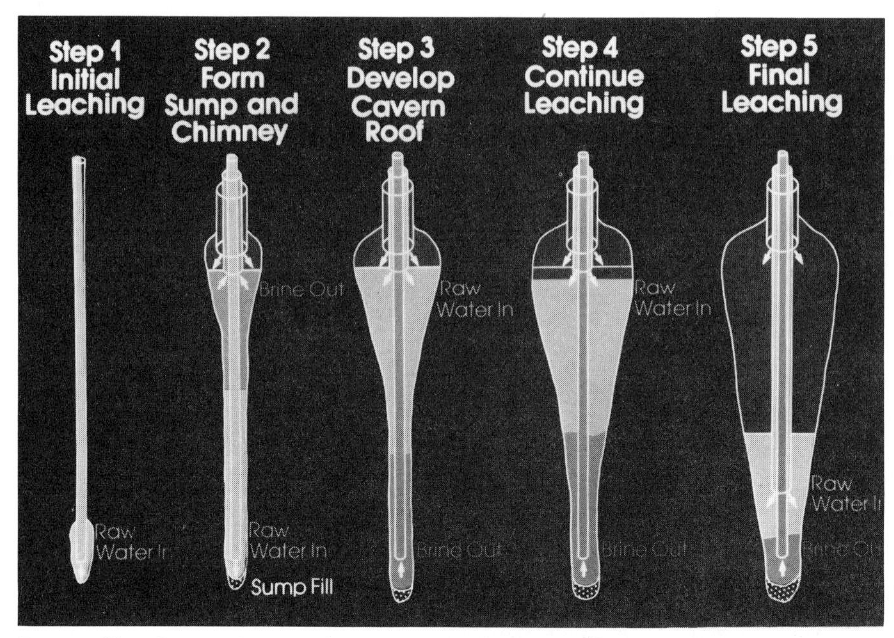

Figure 26. Strategic petroleum reserve, Louisiana/Texas.

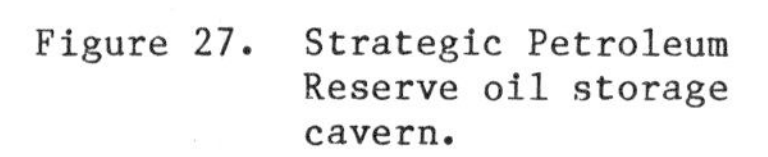
Figure 27. Strategic Petroleum Reserve oil storage cavern.

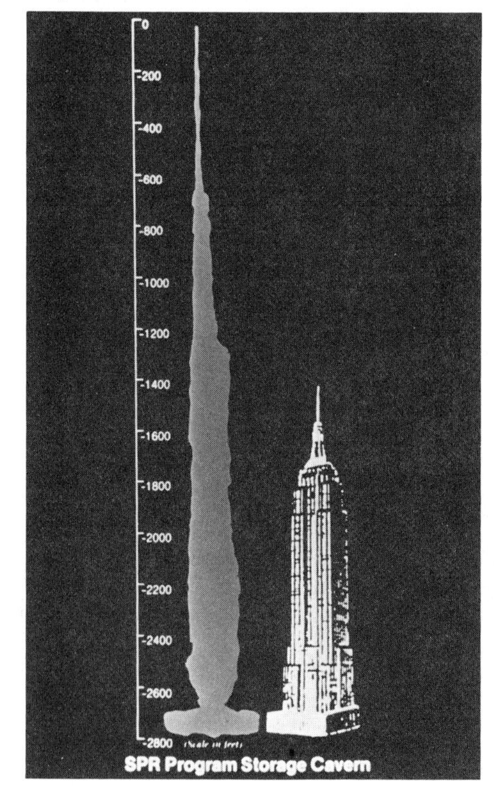

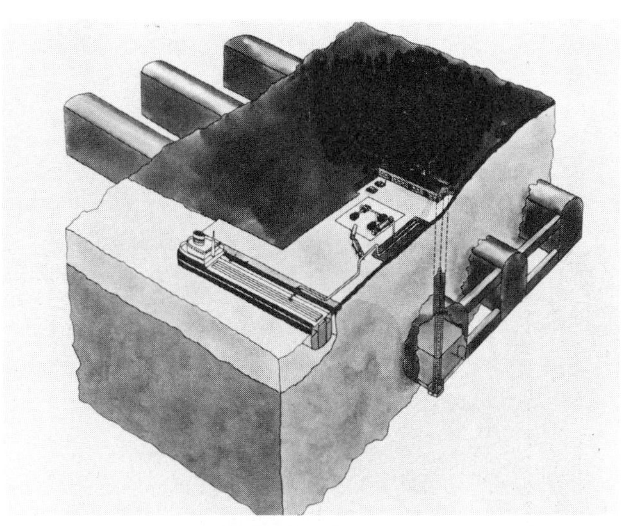

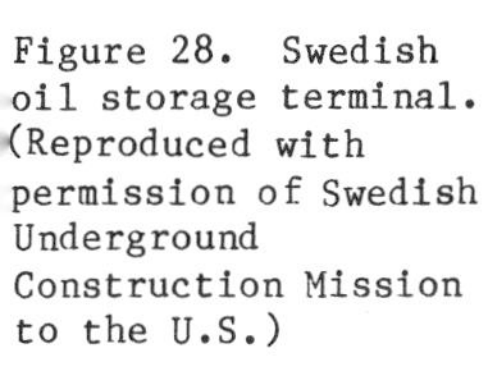

Figure 28. Swedish oil storage terminal. (Reproduced with permission of Swedish Underground Construction Mission to the U.S.)

Figure 29. Swedish oil storage cavern. (Reproduced with permission of Swedish Underground Construction Mission to the U.S.)

Figure 30. Underground housing, Tunisia. (Reproduced with permission of Don Gill, "Underground Space.")

Figure 31. Williamson Hall, University of Minnesota. (Reproduced with permission of Don Gill, "Underground Space.")

Figure 32. Williamson Hall interior. (Reproduced with permission of Don Gill, "Underground Space.")

Figure 33. Underground housing, Minneapolis. (Reproduced with permission of Don Gill, "Underground Space.")

Figure 34. Dune house, Florida. (Reproduced with permission of Great Midwest Corporation.)

Corporation (Fig. 35), which has a major commercial establishment growing on the west side of the city. Figure 36 shows what a limestone mine looks like when you have fixed it up a little bit. The use for storage (Fig. 37) is fairly obvious. But they have gone further than that. Figure 38 shows an underground office center, Chroma-Graphics, a photographic processor where the even temperature and humidity and controlled environment are very conducive to operations. Figure 39 is an underground computer center. Again, the controlled environment is very helpful. Figure 40 is an underground scientific laboratory for measurements of scientific equipment. Now, would you believe that there are underground tennis courts (Fig. 41). I looked for an underground golf course, but I have not found one yet.

Perhaps the largest current activity of tunneling applications is the nuclear waste storage repository program, which has been under discussion for a great many years. They are gradually inching toward doing something about it. A great deal of planning has been done (Fig. 42). Basically the concept is to concentrate the wastes in a series of cannisters which would be buried in the floor of a mined gallery. A series of galleries would be aggregated into a storage panel a quarter of a mile long. A group of panels is in turn aggregated into an underground field of storage caverns with an extraordinarily complex core (Fig. 43), which is required to prosecute excavation of one part of the repository while cannisters are being placed simultaneously in another part, and to provide independent access and ventilation to both parts. There will be several hundred miles of tunnels in one of these underground repositories. The Nuclear Regulatory Commission thinks this is beautiful. They sent copies of Figure 42 to 35 governors saying, "Hey, look what we're going to build in your state. This is beautiful." They seemed to be surprised that they did not get much enthusiasm back. Nevertheless, it is a major problem that we do have to deal with, and it is certainly a macro-project.

I think we have covered the gamut here. This is a survey of the kinds of uses to which tunnels have been and are being put. We now have a fascinating program ahead of us of specific examples.

Figure 35. Great Midwest Corporation, Kansas City. (Reproduced with permission of Great Midwest Corporation.)

Figure 36. Rehabilitated limestone mine, Kansas City. (Reproduced with permission of Great Midwest Corporation.)

Figure 37. Storage in limestone mine. (Reproduced with permission of Great Midwest Corporation.)

Figure 38. Office space in limestone mine. (Reproduced with permission of Great Midwest Corporation.)

Figure 39. Computer center in limestone mine. (Reproduced with permission of Great Midwest Corporation.)

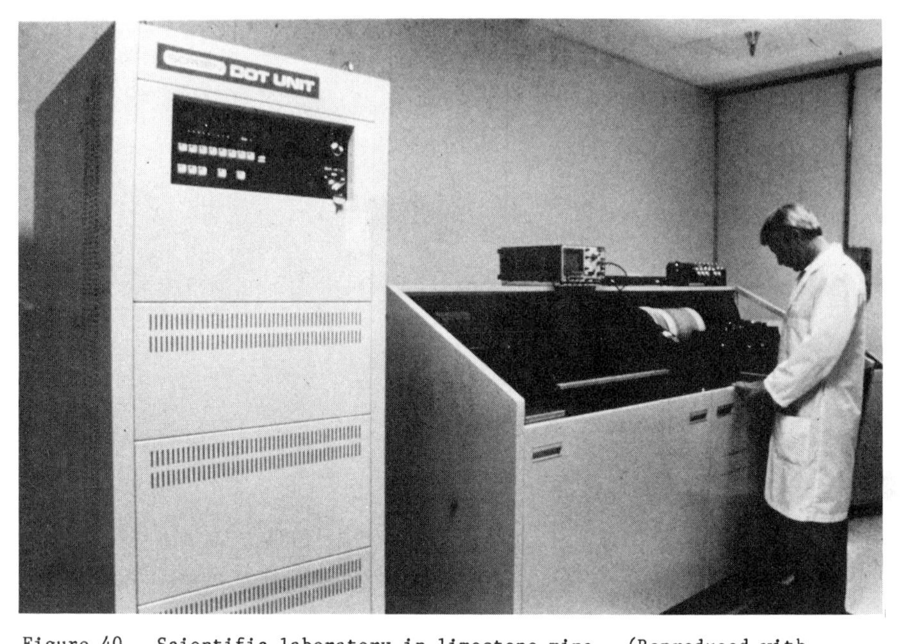

Figure 40. Scientific laboratory in limestone mine. (Reproduced with permission of Great Midwest Corporation.)

Figure 41. Underground tennis court, Carthage, Missouri. (Reproduced with permission of Don Gill, "Underground Space.")

Figure 42. Nuclear waste repository--conceptual layout.

Figure 43. Nuclear waste repository--core complex.

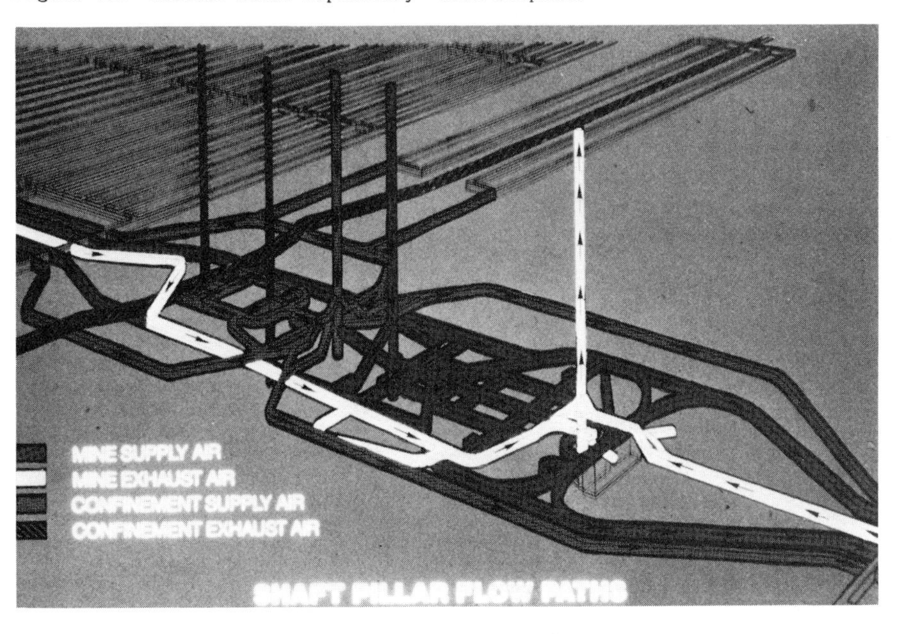

The Economics of Tunneling: Experience and Expectation

SIR ALAN M. MUIR WOOD, Senior Partner
Sir William Halcrow & Partners
Vineyard House
44, Brook Green
London W6 7BY England

In the International Tunnelling Association, and also in the Major Projects Association of the United Kingdom, we are attempting to tackle the problem of what needs to be done to bring professionals together with people involved, at various levels, in the process of decision; the object is to have more successful macro-projects in the future than we have had in the immediate past. As part of a parallel effort, I have been asked today to address the economics of tunneling and to assess both our experience and our expectations. One way to approach the topic is to consider, in an organized manner, the present situation, followed by an inquiry into the prospects and limitations of automation, and concluding with some remarks on future options and tasks.

The Present Situation

First of all, I am concerned with the evolution of a project, because there is a widespread misunderstanding of what we mean by "project costs." In tunneling, when a new project is started, there is no immediate way of ascertaining an expected cost. Let us admit, tunnels start off with expenditures: we are so dependent on the ground that when we begin to study a proposed tunnel, we do not know what we are talking about as far as cost is concerned. Hence, the importance of the evolution of a project during the time when money is spent precisely in order to learn a good deal more about its costs and what the uncertainties of cost are likely to be! In Figure 1, I have endeavored to show in very simple terms how such

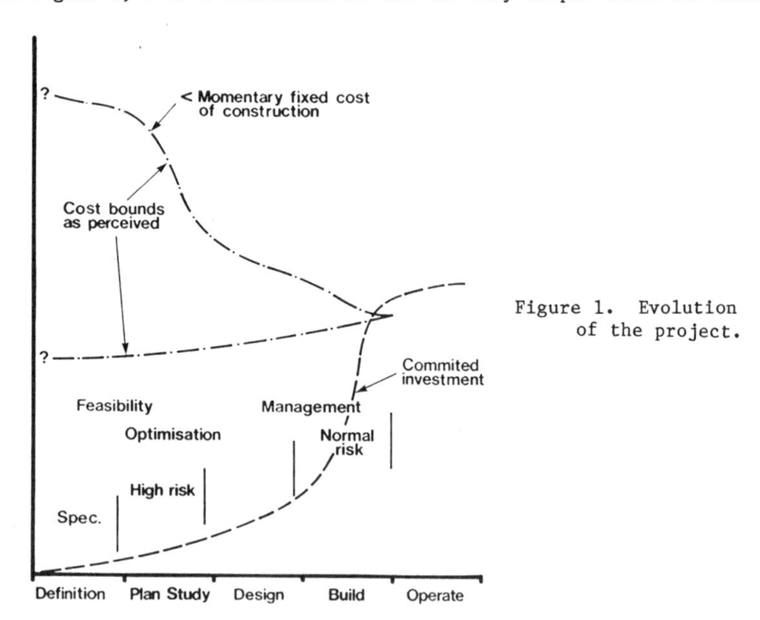

Figure 1. Evolution of the project.

Tunneling and Underground Transport: Future Developments in Technology, Economics, and Policy
F.P. Davidson, Editor

an engineering project evolved, commencing with project definition to find out what it is all about; project planning and study, to reach a state where we are quite settled on what we want to build and at the same time learning more about the conditions in which we shall have to build. The phases of design, building, and operation are parts of an iterative process whose decisions are closely interconnected.

During the early stages, spending a certain amount of money enables us to bring down the boundaries of uncertainty in cost very sharply, enabling us to start the second phase, that of detailed design. Too many people wish to concentrate far too early in a project on the definition of a fixed cost. The problem arose in the period of intergovernmental consideration of competitive bidding for a tunnel or other fixed link across the English Channel. I was concerned myself in preparing some of the technical requirements for those who were to bid on the project. It was difficult, during the few months allowed, for anybody to confidently establish a reasonable fixed cost for this project, bearing in mind the inevitable remaining uncertainties. The period of early study and planning is inherently one of high risk. And so I have designated these three stages as roughly (1) establishing feasibility; (2) establishing project optimization; and (3) management of the project itself once it is defined, during which the financial risk may be defined as moving from "speculative" to "normal for a major project" (Fig. 1).

Knowing a little bit about the evolution in project costs as I have seen them over the past 35 years, I have been struck by the fact that in the 1950-1960 decade, real costs were reduced by a factor of about two (Fig. 2): this was in relation to tunnels built in the easiest circumstances of terrain, where rock and soil conditions were fundamentally favorable. There is an analogy to agricultural history: a "great leap forward" in technology occurred when populations could still exploit a surplus of nearby fertile farmland in the 18th and early 19th centuries, so that improved practice could be transferred rapidly. Now, to grossly over-simplify what has happened in tunneling, I would say that

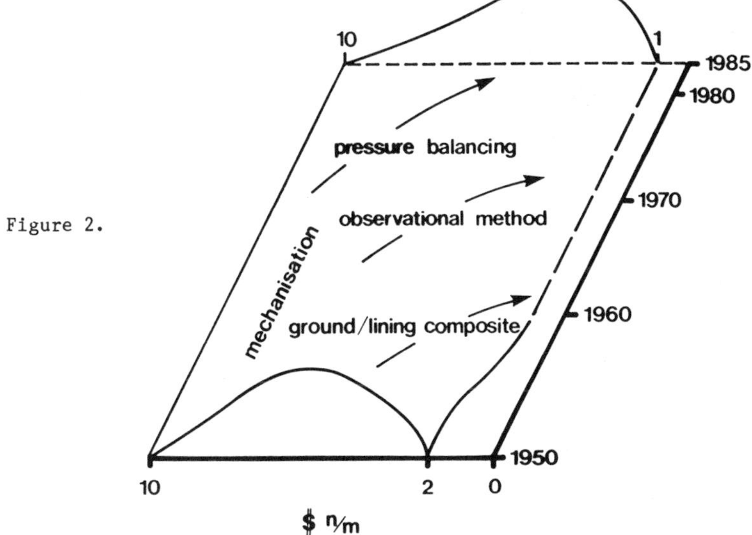

Figure 2.

the next real step forward was made when we made much wider use of the observational method. The observational method implies that we accept the idea that we do not understand a great deal about the very complex behavior of the ground in which we are building the tunnel. We develop fairly simple theories of how the ground is going to behave; as we carry forward the construction process, we then measure expectation against what is actually happening.

But to make such "second grade corrections" with minimal disruption, there has to be a very direct relationship between design and construction (Fig. 3). If we're trying to design a tunnel but then stop it all and hand it over to somebody else who is starting from scratch, there is a full risk at that point and you cannot make use of the observational method. This is one of the principal reasons for understanding the nature of the iterative process, that is, the successive steps that are indispensable and interactive--what we are really doing--and how the various aspects of building a tunnel are related one to the other.

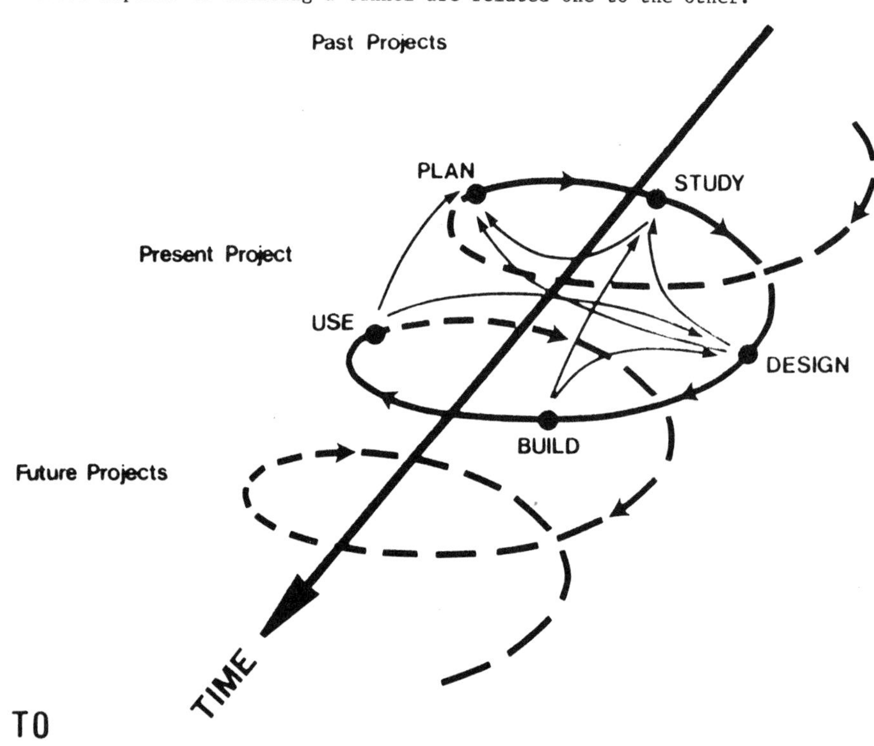

Figure 3. Tunneling: The interactive nature of learning.

During the last 15 years or so, what I would describe as a major evolution has come in what I call "pressure balancing." This technology involves a whole variety of shields used in ground which is not stable. Referring again to Figure 2, it is possible to envisage how, during a period of 35 years, we started off by reducing the costs of the tunnels that were easiest--and cheapest--to dig. The observational method then began to extend this progress to the medium cost projects; and finally pressure balancing developments have improved our ability to cope with the more expensive tunnels. This, I think, is a generally valid framework for understanding cost changes in the past, greatly oversimplified of course, and I think it gives us some idea about the problem of predicting future developments. Of course, we have to be concerned with the use of the project, and this brings us right back full circle to the planning and design.

With this as a model, which I see we should be trying to work toward, I have to confront the question of why it is that we are not always successful in what we are trying to do. When I say "we," I am speaking for a fairly wide community of people who I believe feel as strongly as I do in this respect. We start off with the point that if we are capable in a particular project of applying what I call the ideal technology, what does this imply? First of all, unless there is somebody who can actually see the project all the way through its various phases and relate one to the other, then I believe we cannot get anywhere near optimum cost. If there is going to be a division between planning on the one hand, design on the other, ground investigation, construction on another, with full responsibility taken over and transferred during these phases, I believe that this is disastrous as far as economic tunneling is concerned.

Ideal technology entails continuity through the planning, study, and inquiry phases; the design process (which must continue through the period of construction); and construction itself. The whole process needs to be related to operation and maintenance to enable decisions to be taken on WHOLE LIFE COSTS. It also implies a thorough understanding of the GEOLOGY and its consequences in terms of planning, design and construction. And there must be an appropriate contractual structure.

Barriers to the ideal include the following three factors: (1) incompetence of any of the major parties concerned: an incompetent owner may lead to the appointment of an incompetent engineer who, in turn, cannot recognize incompetence in potential contractors; (2) discontinuity, which makes incremental improvement and also the recovery of costs of major improvement very difficult indeed; and (3) inappropriate contractual arrangements which exacerbate the transfer of responsibility between parties, especially where restraint and cohesion are needed in situations of uncertainty. (In the United States, for example, it is common practice for a contractor to accept full responsibility for a project--including its design--while geology remains full of surprises.) Perhaps such procedures led to the old definition of a tunnel as "a long cylindrical hole in a state of plane strain with a geologist at one end and two lawyers at the other."

If we take unit cost for ideal technology for a given type and size of tunnel as a range of 1 to 10 (see Fig. 2), then for situations in which the several barriers occur (and they tend to occur collectively since all spring from attempts for least cost rather than best value for money), the range may be represented as 2 to 50+.

Means towards the ideal entail the following factors:

- TECHNICAL AUDIT of practices and organizations concerned with all aspects of tunneling to determine whether systems and qualitative capabilities lead to best value, as against FINANCIAL AUDIT which simply looks at cost and fiduciary performance. (My impression is that VALUE ENGINEERING as currently practiced is usually introduced too late in the evolution of a project to achieve best value for money and only transfers part of that benefit to the owner!)

- GEOTECHNICAL COMPETENCE should ensure that geologists, geo-engineers, and tunnelers are properly linked, so that information relevant to economic tunneling and to the determination of problems is identified. Too often the process is geologist led and produces much interesting information--but not for the tunnel project.

- Finally, regard must be given to the ITERATIVE nature of LEARNING. This not only affects linkages for a single project (e.g., between stages of investigation and design; affecting design during construction) but the understanding carried forward from past to present projects, much of it implicit and not consciously perceived in making ENGINEER JUDGMENTS (see Figure 3).

For ideal technology (as defined above--it is by no means the norm in any country), the recent evolution in unit costs in real terms is simplified as below:

- In 1950, for the conditions normally accepted for tunneling, using the STATE OF CAPABILITY of the time, the range of costs is <u>2 to 10</u>, mean <u>6</u>.

- Between 1950 and 1960, the many developments that enabled design to be based on COMPOSITE ACTION between ground and lining (or support) affected tunneling in the optimal ground conditions, the range of costs then changed to <u>1 to 10</u>, mean <u>4</u>.

- Since 1960, wider adoption of the OBSERVATIONAL METHOD, while not affecting the range, reduced the mean.

- Since 1970, PRESSURE BALANCING has reduced the cost of much tunneling in weak waterbearing ground, so the range may be expressed as, say, <u>1 to 8</u>, mean <u>3</u>.

- Throughout the period, increasing MECHANIZATION, particularly the development of TBMs and roadheader-type machines, has assisted in reducing mean costs. (All the figures against the above statements should be seen as indicative only, and are essentially subjective.)

<u>Evolving Economics of a Project</u>

As a project develops through its phases of definition, design, and execution (see Fig. 1), final cost uncertainty is reduced (assuming litigation is avoided). The process can be (very roughly) tabulated, as shown in Table 1.

TABLE 1. Principal Aspects of Development

STAGE	FINANCIAL RISK	COMMITTED INVESTMENT (%)	RATIO OF BOUNDS OF PERCEIVED COST
(a) Definition	Speculative	0-0.5	10 reducing
(b) Planning	High risk	0.5-2	5 reducing
(c) Design	Intermediate risk	2-4	2 reducing
(d) Construction	Normal risk	4-100	1.2 reducing to 1
(e) Operation	Predictable	100	1

Note:

FEASIBILITY stages (a) and part (b)
OPTIMIZATION stages (b) and (c)
MANAGEMENT stages (c), (d) and (e)

Automation: Prospects and Limitations

Already a great deal is being done in tunnel surveying to control line and level. This enables us to anticipate what is likely to happen, particularly when we are controlling shields and tunnel boring machines, so that we can make corrections before they are necessary. When we are operating tunnel boring machines, we can already begin to control the thrust of the torque and then the rotation, and so forth, by measurements coming through the monitoring of the machine. We can also do a great deal to control drilling patterns for drill and blast methods of tunneling, and to control road headers. Comparable progress is being made with the earth balance machines.

In the future, I believe that one of the important things is that automation will enable us to monitor machines in a manner that will reduce breakdowns and therefore "down time." But there are limits to automation, because of the subtleties of the ground. We still have to be prepared to react to changes in the ground, and for this I think we must depend on people. I do not see expert systems here coming to replace the tunnel engineer in that respect. On the other hand, we have to ensure that our tunneling machine will be infinitely tolerant of possible changes in the ground. This has a great deal to do with economics. Information technology is already giving us a much better common project data source used by all parties to a project so that we have all the information put together in such a way that it can be more readily handled. Those of us with personal tunneling experience know that for the future we have to devise methods that will enable others to use our experience and the experience of our colleagues.

One area where improvement can be anticipated is in increasing the reliability of our machinery. Despite all the inventions and innovations of recent years, downtime still represents close to 80% (see Fig. 4). This means that the figure of usage over a period of years is about one-fifth of what you would find as a peak rate. The right approach is to

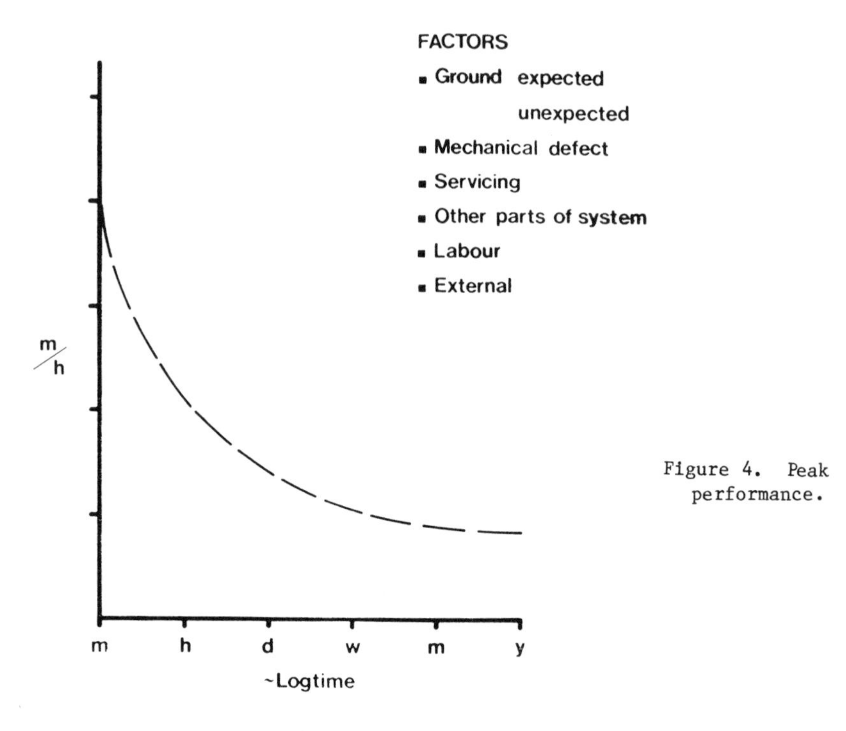

Figure 4. Peak performance.

view the whole process as a system. By insisting upon quality performance standards at vital points of the system, we shall gradually upgrade reliability, and therefore output.

The main causes of delay may be categorized: (1) ground variation (expected and unexpected); (2) mechanical/electrical/electronic defects and their servicing; (3) defect or incapacity of other components of system management, including labor; and (4) external factors and problems.

Future Options and Tasks

A fully automated tunneling system infinitely tolerant to ground variability is an illusion. But the mining industry, with companies habitually digging 500 or more miles of tunnels per year per company, has both the incentive and the capital to explore promising new methods and machines, and I am sure that further progress toward automation will be made within the next few years. The main point is that without first attending to the essential elements for improved tunneling practices, including the contractual setting, we are not going to see the order-of-magnitude improvements potentially achievable in many circumstances.

I have been intrigued by the presentation of a futuristic transport system called "Planetran," with its vision of supersonic levitated trains hurtling through evacuated tunnels or tubes under land and sea. Planetran depends for its practicality on many factors. I will ignore the enormous problem of financing a project which is incapable of any return before the sinking of several billion dollars. I will address the principal problem in civil engineering terms. With the appropriate approach, I see no problem in meeting stringent alignment demands; these will be of the same order as the LEP electron-positron collider at CERN. The main problems, therefore, lie in speed and cost of construction and standards for sealing.

For the greater part, the routes would have considerable tolerance in routing, respecting sites for terminal and intermediate stations, at intervals of hundreds of miles, and acceptable curvatures.

The most essential requirement is a thorough ground exploration, particularly in relation to structure and tectonics. We can now achieve spectacular results with geophysical profiling and its derivatives, a "quantum leap" from the "sparker" days. There are types of ground to be favored, types to be tolerated, and types to be avoided at all reasonable costs.

For some of the current projects which will involve new forms of application of tunnels, studies are being made of the means which will permit average rates of advance to represent a higher fraction of peak rates (see Fig. 4). This is one of the examples of "system engineering" to be brought to bear on the project. Cost depends partly on the ground, partly on the choice of system, partly on the rate of advance.

Sealing is the most difficult of all the problems, considering the great volume to be evacuated and the standards to be achieved. The most challenging design problem is how to attain this end without imposing an uneconomical degree of loading upon the tunnel. I can perceive one or two strategems for cracking this challenge, but these need to be considered against the appropriate range of ground and environmental standards.

The concept should not be dismissed as an idle fantasy. A desk study could certainly test out the practicality of certain alternative approaches and help create a sharper definition of outstanding problem areas. We live in a world that is increasingly becoming an interlinked neighborhood; the art of tunneling remains a substantial resource. If progress is incremental--and intermittent--it is progress nonetheless. And attention to essentials will both improve the record of the tunneling industry and help justify those larger hopes and expectations which are characteristic of the human story.

Construction of the Seikan Tunnel

MINORU SHIMOKAWACHI
Superintendent, Yoshioka Construction Factory,
Seikan Construction Bureau, JRPCC, Hokkaido, Japan

INTRODUCTION

Japan consists of four main islands, as shown in Figure 1. Kyushu is connected with Honshu by four fixed links: two railway tunnels for conventional and Shinkansen trains, the road tunnel, and a suspension bridge carrying an express highway.

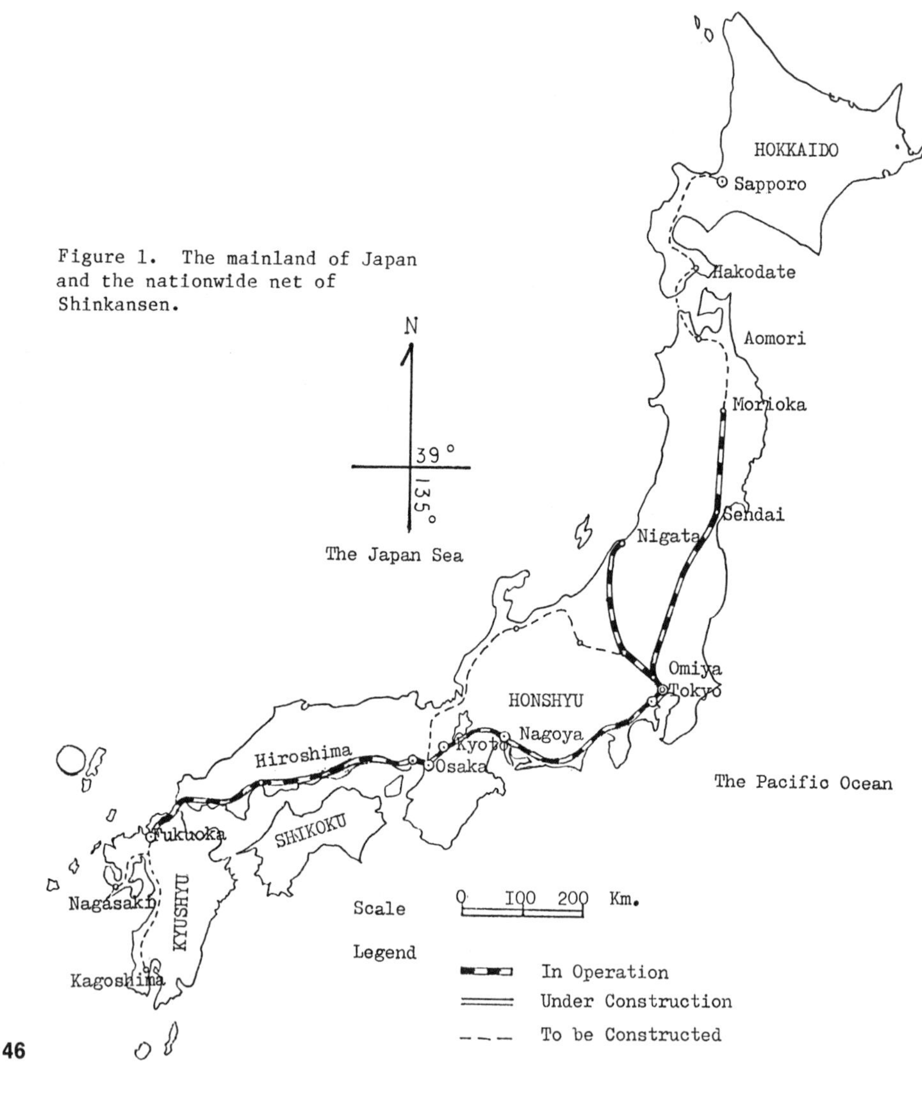

Figure 1. The mainland of Japan and the nationwide net of Shinkansen.

Tunneling and Underground Transport: Future Developments in Technology, Economics, and Policy
F.P. Davidson, Editor

Two gigantic projects are now under construction:

-- The undersea tunnel project between Hokkaido and Honshu; and
-- The suspension bridges between Shikoku and Honshu.

The history of the concept of the Seikan Undersea Tunnel can be traced back to around 1918. Japanese National Railways (JNR) engaged in intensive investigations from 1946 to 1964, when the Japan Railway Construction Public Corporation (JRPCC) was founded. At that time, responsibility for the project was handed over by JNR to JRPCC. JRPCC continued the work of exploratory tunnel excavation at Yoshioka on Hokkaido and at Tappi on the Honshu side.

The alignment of the tunnel was chosen to cross the western side of the Tsugaru Straits as a result of detailed studies of geology, topography, and seismic aspects of the Straits.

Through multi-disciplinary investigations, the data and information necessary to decide upon a specific route were obtained, and the techniques of exploratory boring, grouting, and shotcreting were developed into practical methods. The feasibility of the tunnel project was confirmed. The final plan, shown in Figure 2, was authorized and full-scale construction commenced in 1971.

PLANNING

The Seikan Tunnel was designed for the combined use of both Shinkansen and conventional railways; its main characteristics are shown in Table 1.

TABLE 1. The Standard Design of the Main Tunnel.

Maximum gradient		I2 0/00	
Plane	Simple curve radius	6500 m	8000 m
Curve	Transition length	215 m	165 m
Profile	Minimum thickness from sea bottom		100 m
	Maximum depth of sea		140 m
Cross section of tunnel		Shinkansen double track	

The standard cross section, as shown in Figure 3, is adopted where favorable ground conditions permit, while the circular or thick side-wall types are employed in fractured or loose-ground conditions. The topography of the seabed along the alignment is shown in Figure 4: the tunnel passes under the saddle portion of the sea bottom. The maximum sea depth is 140 meters. The tunnel alignment was decided through careful consideration of topography and geology. Particular attention was paid to the existence of the larger faults.

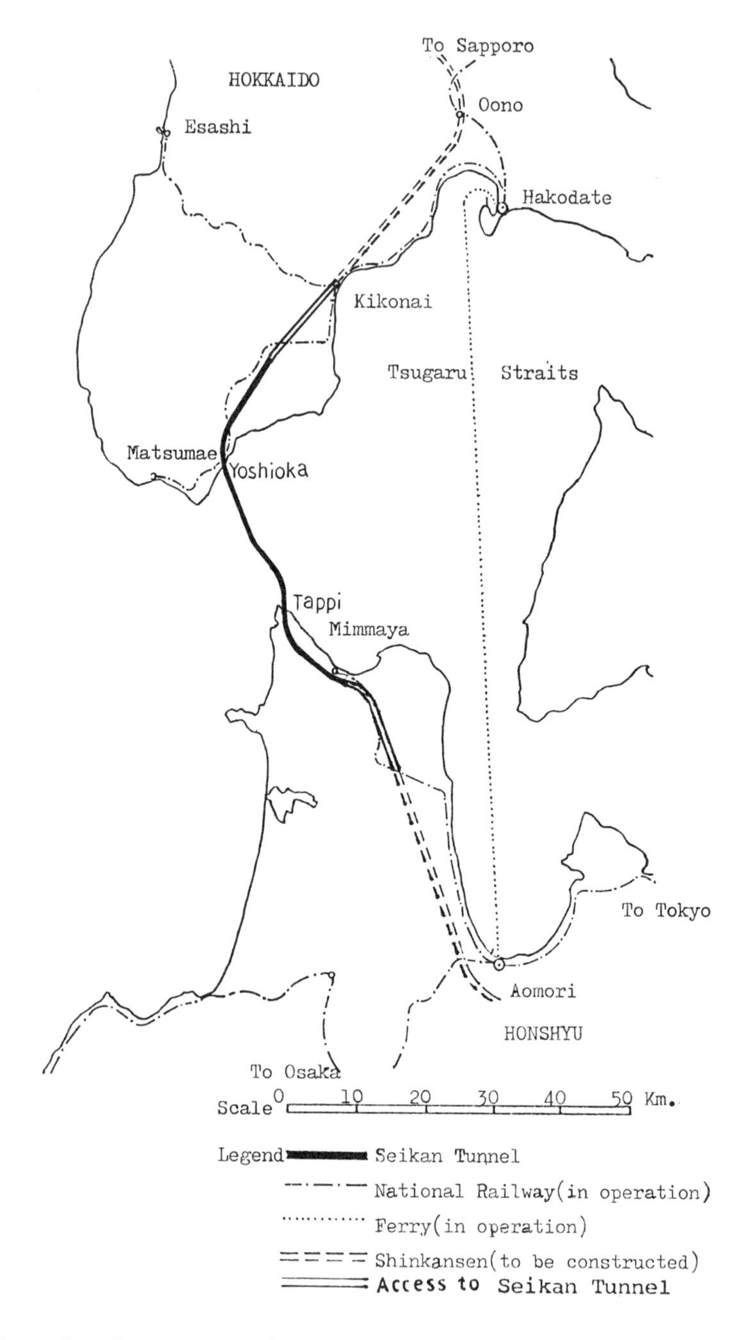

Figure 2. The vicinity of the west entrance of the Tsugaru Straits and the net of railways.

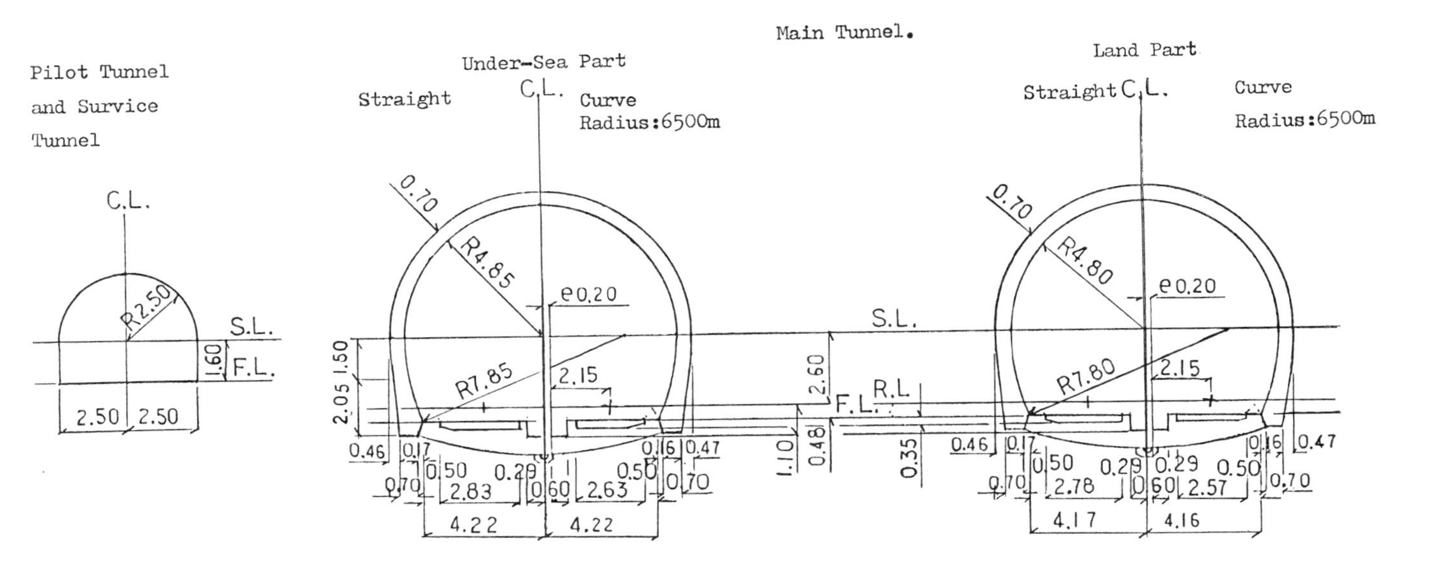

Figure 3. The standard cross section of the Seikan Tunnel.
(Unit length: meter)

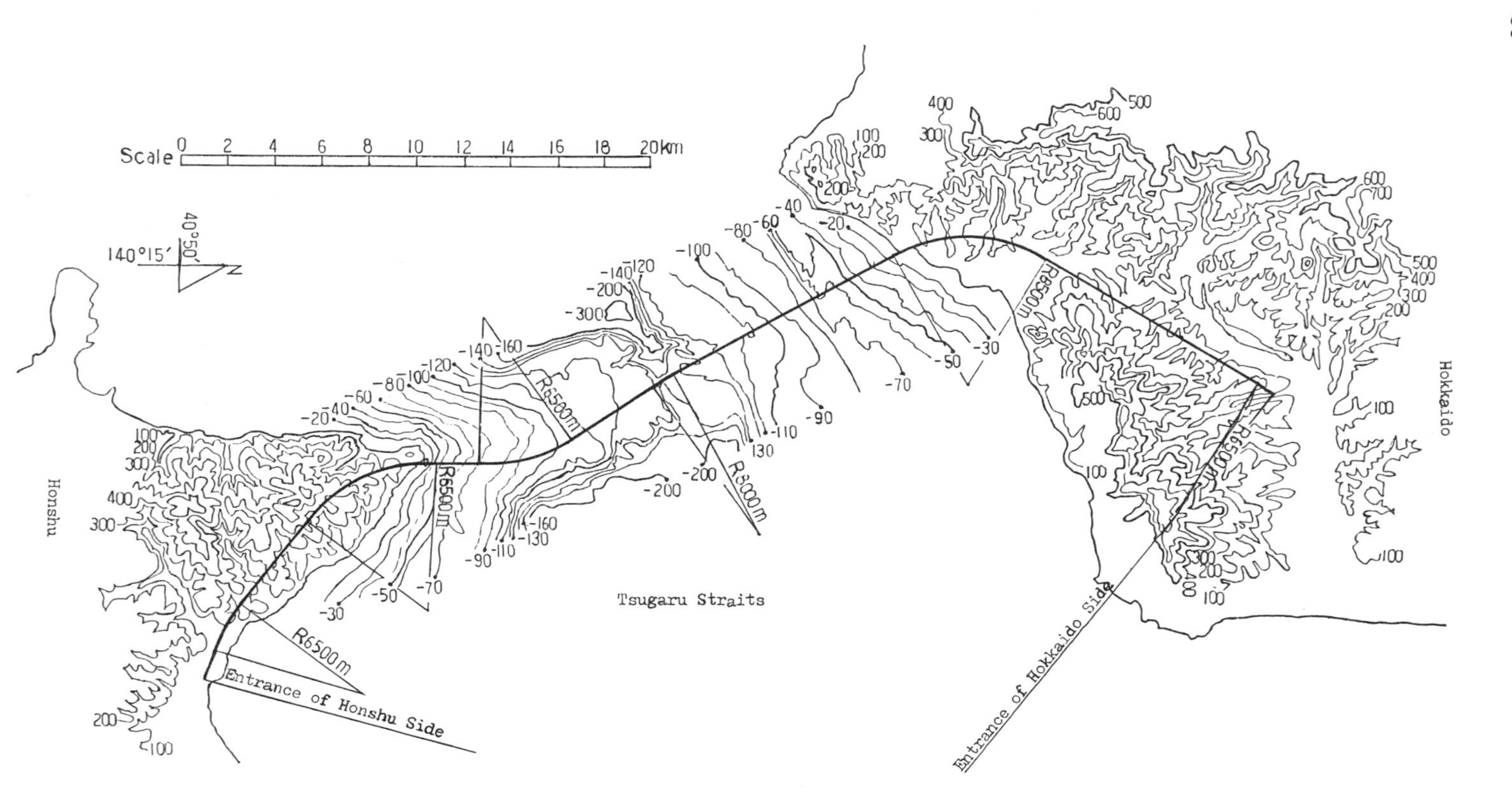

Figure 4. The topography of the west entrance of the Tsugaru Straits and the route of the Seikan Tunnel. (Elevation unit: meter)

The plan and longitudinal profile are shown in Figure 5. The undersea part is so long that the pilot and service tunnels are excavated along the main tunnel in order to be utilized as ventilation, drainage, and transportation galleries during both the stages of (1) construction and (2) the installation of services after completion.

The pilot tunnel must be advanced first to survey the distribution and character of geology and ground water, against which the more suitable methods are evaluated and selected.

The service tunnel must be advanced following the pilot, then the data in both tunnels are compared to ascertain which methods can best be utilized for the main tunnel.

CONSTRUCTION

According to the conditions of geology and topography, the total length of 53 kilometers was divided into nine construction sections, as shown in Figure 6. Tappi and Yoshioka sections have the pilot and service tunnel branches from the inclined shaft, and also from the vertical shaft, respectively.

The service tunnel and the pilot tunnel converge together along the 5 kilometer-long section of the sea portion, where the pilot tunnel plays the twin roles of service and pilot gallery.

The main tunnel is excavated from the service tunnel through the connecting galleries which are situated at about 600-meter intervals. Normally, the main tunnel under the sea section is excavated simultaneously at three different faces on both sides of Honshu and Hokkaido.

The configuration of the pilot, service, and main tunnels is explained in Figure 7.

In the pilot tunnel, the water at the face is drained behind by the portable pumps and flows naturally to the bottom of the inclined shaft in which the water is pumped up and out by stationary pumps. Water in the service tunnel is drained by portable pumps to the nearest ventilation shaft, through which the water flows to the pilot tunnel. Water in the main tunnel is drained by the portable pumps to the nearest gallery, through which the drainage flows to the service tunnel.

The ventilation system of the undersea section during construction is characterized by the employment of each tunnel and shaft as the main inlet and outlet of air flow.

Contaminated air is blown up through the vertical shaft by stationary fans; this induces the fresh air to come in through the inclined shaft. The fresh air circulates, in turn, through pilot tunnel, service tunnel, and main tunnel. The service tunnel is connected with the pilot tunnel by vertical ventilation shafts.

In order to secure smooth circulation of air flow, the air locks of the main tunnel are installed at the border of the sea and land section. Portable fans and ducts are used for ventilation of the blind working sites such as the tunneling face.

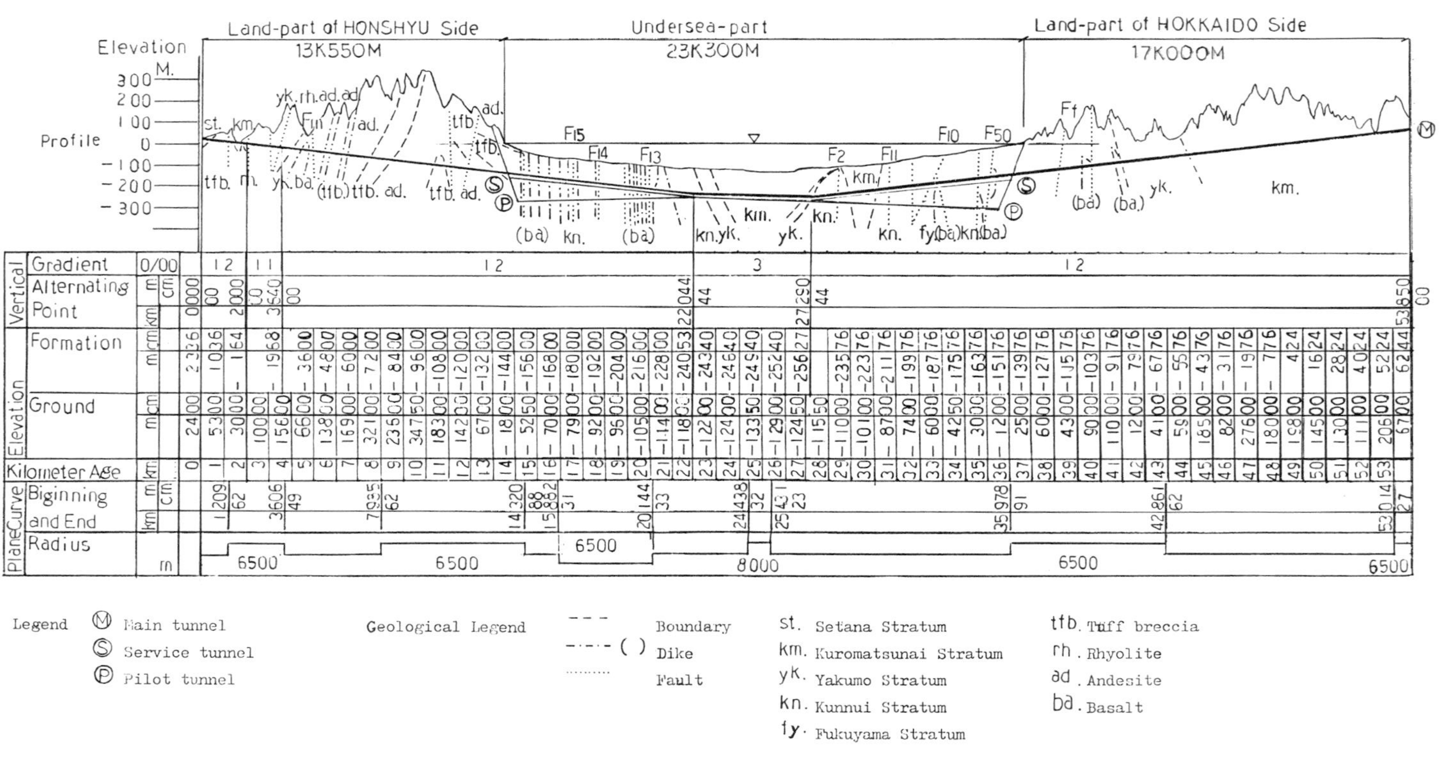

Figure 5. The profile and geology of the Seikan Tunnel.

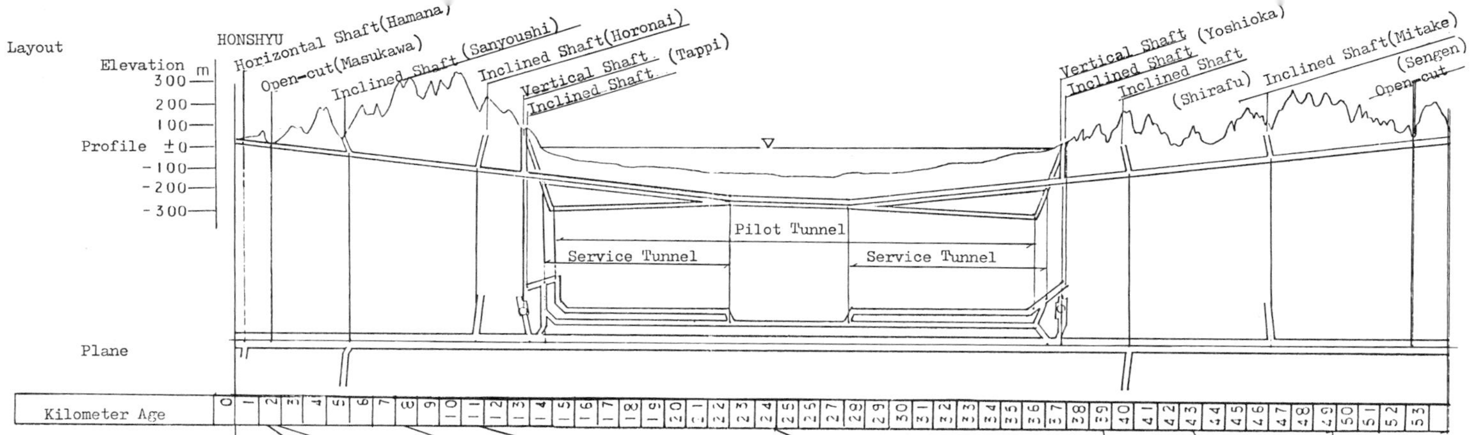

Layout

HONSHYU

Horizontal Shaft(Hamana)
Open-cut(Masukawa)
Inclined Shaft (Sanyoushi)
Inclined Shaft(Horonai)
Vertical Shaft (Tappi)
Inclined Shaft
Vertical Shaft (Yoshioka)
Inclined Shaft
Inclined Shaft (Shirafu)
Inclined Shaft(Mitake)
(Sengen)
Open-cut

Elevation m: 300, 200, 100, Profile ±0, -100, -200, -300

Pilot Tunnel

Service Tunnel

Service Tunnel

Plane

Kilometer Age	0	1	2	3	4	5	6	7	8	9	10	11	12	13	14	15	16	17	18	19	20	21	22	23	24	25	26	27	28	29	30	31	32	33	34	35	36	37	38	39	40	41	42	43	44	45	46	47	48	49	50	51	52	53

				Hamana	Masukawa	Sanyoushi	Horonai	Tappi		Yoshioka		Shirafu	Mitake	Sengen
Section	Name			Hamana	Masukawa	Sanyoushi	Horonai	Tappi		Yoshioka		Shirafu	Mitake	Sengen
	Biginning and End		m 000	1 470	1 908	7 400	10 900	23 900		38 600		42 500	48 900	53 850
			km 0	1	1	7	10	23		38		42	48	53
	Length		m	I470	438	5492	3500	I3000		I4700		3900	6400	4950
Accessing to Main Tunnel	Juncture			0K394m	IK540m	4K920m	I0K630m	I3K000m	I3K639m	36K000m	36K575m	34K490m	45K750m	52K270m
	Horizontal Part		m	6Im000	I20mab.	I35mI75	I67m793	I95m390	594m863	294m330	270m395	84m4I0	I72m700	I5mab.
	Vertical Shaft		m					I78m500	↕	↕	I67mII0			
	Inclined Shaft		m			276m4I6	584m4000		760m000	669m5I0		495m520	429m790	
Facilities	Muck	Conveying		Trolley	Trolley	Trolley	Trolley	Trolley		Trolley		Trolley	Trolley	Trolley
		Carring out		Trolley	Bucket	Belt-Con.	Belt-Con.	Belt-Conveyer		Belt-Conveyer		Belt-Conveyer	Belt-Conveyer	Bucket-Elevator
	Concrete	Mixing		Out-Plant	Ready-Mix	Out-Plant	Out-Plant	Out-Plant, In-Plant		In-Plant	Out-Plant	Out-Plant	Out-plant	Out-Plant
		Carring in		Agi.Car	Shooter	ShootPipe	ShootPipe	ShootePipe			ShootPipe	Belt-Conveyer	ShootPipe	Shooter
	Elec. Power	Receptacle	kva	3000		5000		23000			I8000	5000	4500	4500
		Generator		300		625	I250	I7250			I3875	625	625	500

Figure 6. The outline of the construction of the Seikan Tunnel.

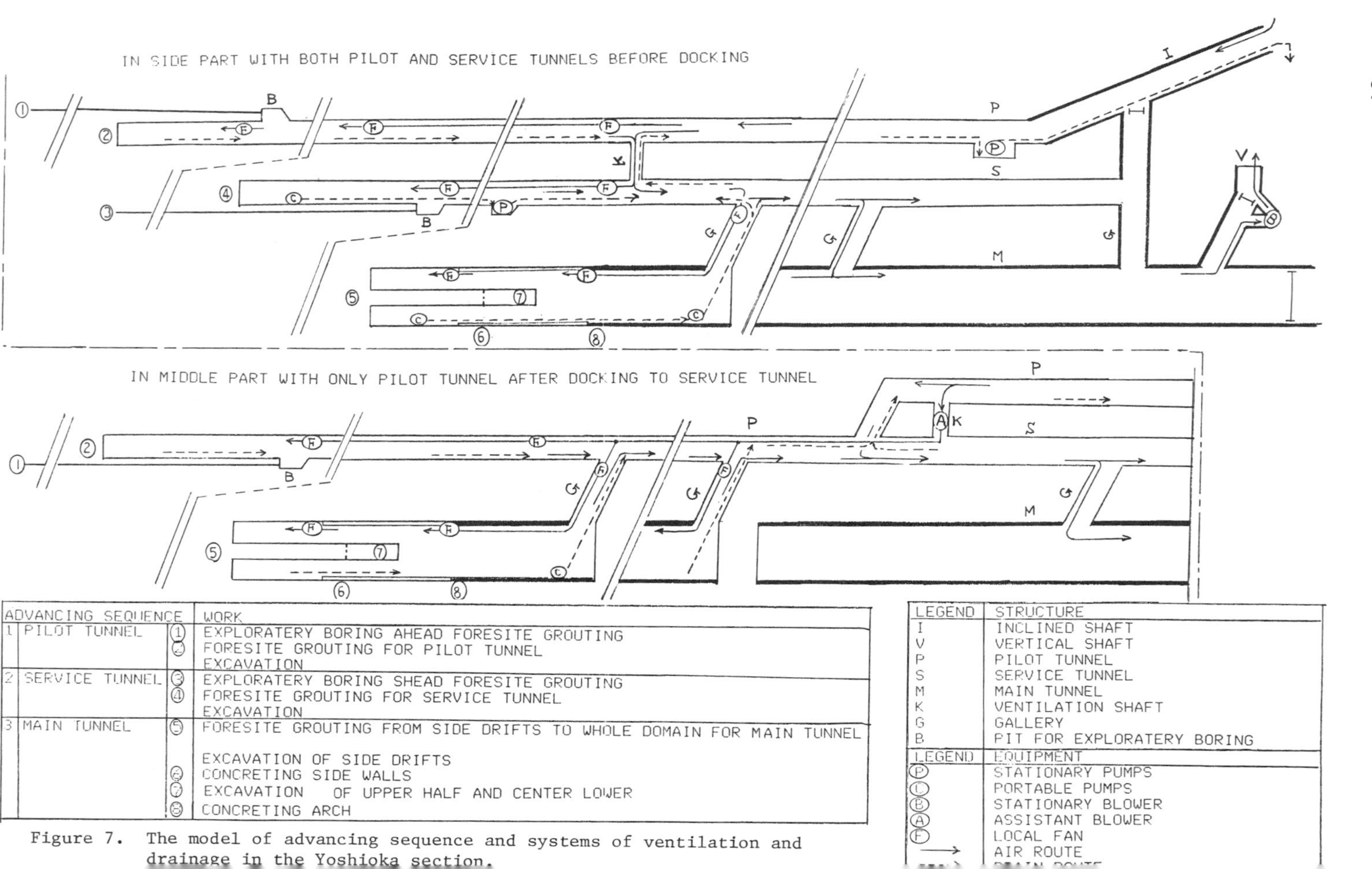

ADVANCING SEQUENCE		WORK
1 PILOT TUNNEL	①	EXPLORATERY BORING AHEAD FORESITE GROUTING
	②	FORESITE GROUTING FOR PILOT TUNNEL EXCAVATION
2 SERVICE TUNNEL	③	EXPLORATERY BORING SHEAD FORESITE GROUTING
	④	FORESITE GROUTING FOR SERVICE TUNNEL EXCAVATION
3 MAIN TUNNEL	⑤	FORESITE GROUTING FROM SIDE DRIFTS TO WHOLE DOMAIN FOR MAIN TUNNEL
		EXCAVATION OF SIDE DRIFTS
	⑥	CONCRETING SIDE WALLS
	⑦	EXCAVATION OF UPPER HALF AND CENTER LOWER
	⑧	CONCRETING ARCH

LEGEND	STRUCTURE
I	INCLINED SHAFT
V	VERTICAL SHAFT
P	PILOT TUNNEL
S	SERVICE TUNNEL
M	MAIN TUNNEL
K	VENTILATION SHAFT
G	GALLERY
B	PIT FOR EXPLORATERY BORING

LEGEND	EQUIPMENT
Ⓟ	STATIONARY PUMPS
Ⓒ	PORTABLE PUMPS
Ⓑ	STATIONARY BLOWER
Ⓐ	ASSISTANT BLOWER
Ⓕ	LOCAL FAN
⟶	AIR ROUTE

Figure 7. The model of advancing sequence and systems of ventilation and drainage in the Yoshioka section.

EXPLORATORY BORING

Information from the offshore investigations was not sufficient to plan tunneling practically under the sea, with the consequence that exploratory boring is usually executed ahead of the pilot and the service tunnel in order to confirm the state of the geology and ground water before excavation.

In order to carry out exploratory boring while the tunnel excavation continues, the boring pit was installed beside the tunnel in a zigzag pattern and the boring and the excavation advanced as shown in Figure 8 and Figure 9. Of course, three or more borings were executed in the section where the geology is very complicated.

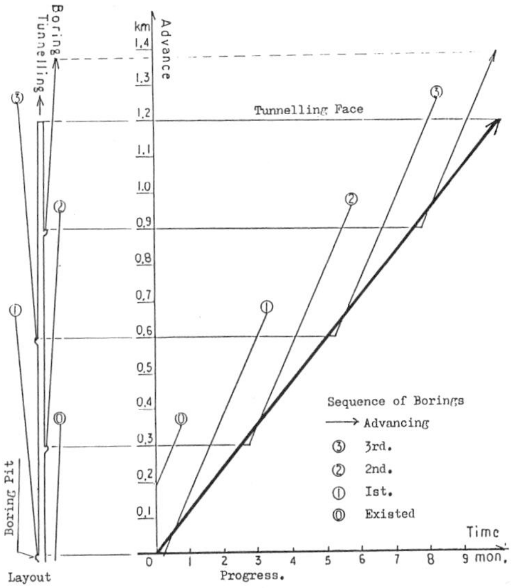

Figure 8. A model pattern of exploratory boring and tunneling.

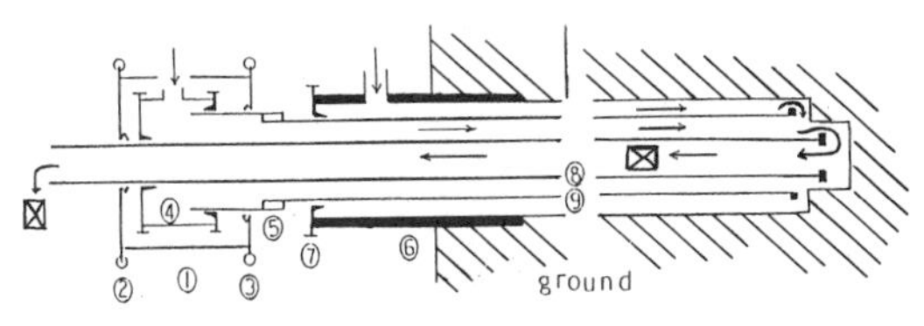

Figure 9. The reverse circulation method using double boring pipes.

NO.	MEMBERS
1	SPINDLE
2	CHUCKS TO INNER-PIPE
3	CHUCKS TO OUTER-PIPE
4	PREVENTER OF ORIGINAL PIPE
5	CONNECTER TO OUTER-PIPE
6	MOUTH-PIPE
7	PREVENTER OF MOUTH-PIPE
8	INNER-PIPE
9	OUTER-PIPE
→	WATER CIRCURATION
⊠	CORE OR CUTTINGS

Horizontal exploratory boring has difficult problems which are not common to vertical boring. These are: deviation from target direction, clogging at the front cuttings, and collapse of the boring hole, mainly caused by the friction between the boring rod and uneven ground layers. In order to overcome those obstacles and increase boring length and speed, the double-casing reverse circulation method has been developed. The maximum record is 2150 meters.

GROUTING

The grouted zone is shaped in a cylindrical form along the tunneling axis. The cylindrical radius, which is variable with conditions of geology and water inflow estimation, is 3 to 5 times the tunnel radius. Figure 10 shows a typical model of the grouting pattern.

Because the ground water has nearly the same hydraulic head as the depth from sea level and similar components to sea water, it is desirable to keep the grouted zone permanently tight. The grouting materials are patented by JRCC. The strength of the hardened materials and setting time are variable with the density of the grouted materials.

The grouting rate and pressure and the density of materials with grouting time are selected in the grouting modes which are classified to the many cases with the geology and ground water.

MAIN TUNNEL

Grouting is used in the main tunnel as well. It is usual, in the main tunnel, that the ground covering all the cross section to be excavated is grouted from the heading face (for example, faces of side drifts in the side drift heading method). But it is often in the great fractured zone that the grouting from the service or the pilot tunnel is used together from the heading face. The grouting in the main tunnel has mainly the same patterns, modes, and materials as in the pilot or service tunnel, but not the same in detail because the main tunnel has about five times the diameter of the pilot tunnel. After the ground is grouted, the ordinary mountain methods are adopted for excavation, as shown in Figure 11.

The bottom drift heading method is used only in the very good ground or vicinity of juncture to the gallery, but the side drifts heading is used for most of the undersea parts--in swelling or loose ground, the side drift (double lining) method is used to set the type of thick side wall on the cross section. In worse ground, an upper half advancing method is used to construct the circular type cross section.

The stiffened support (for example, hooped steel pipe), shotcrete, and rock bolt are used to act together soon after excavation in bad geology, in order to prevent the grouted ground around the excavation from loosening; this is usually monitored with measurements (extensometer, convergence-meter, etc.). Of course, it is usual to shotcrete and rockbolt the weak portion in the common ground.

PRESENT STATUS

The excavation of the pilot tunnel was finished in January, 1983. The undersea portion of the main tunnel has been virtually completed.

Now we are progressively constructing the facilities and machinery for operation in the Seikan Tunnel. We have started to set up the equipment and electric facilities to drive trains into the main tunnel, as shown in

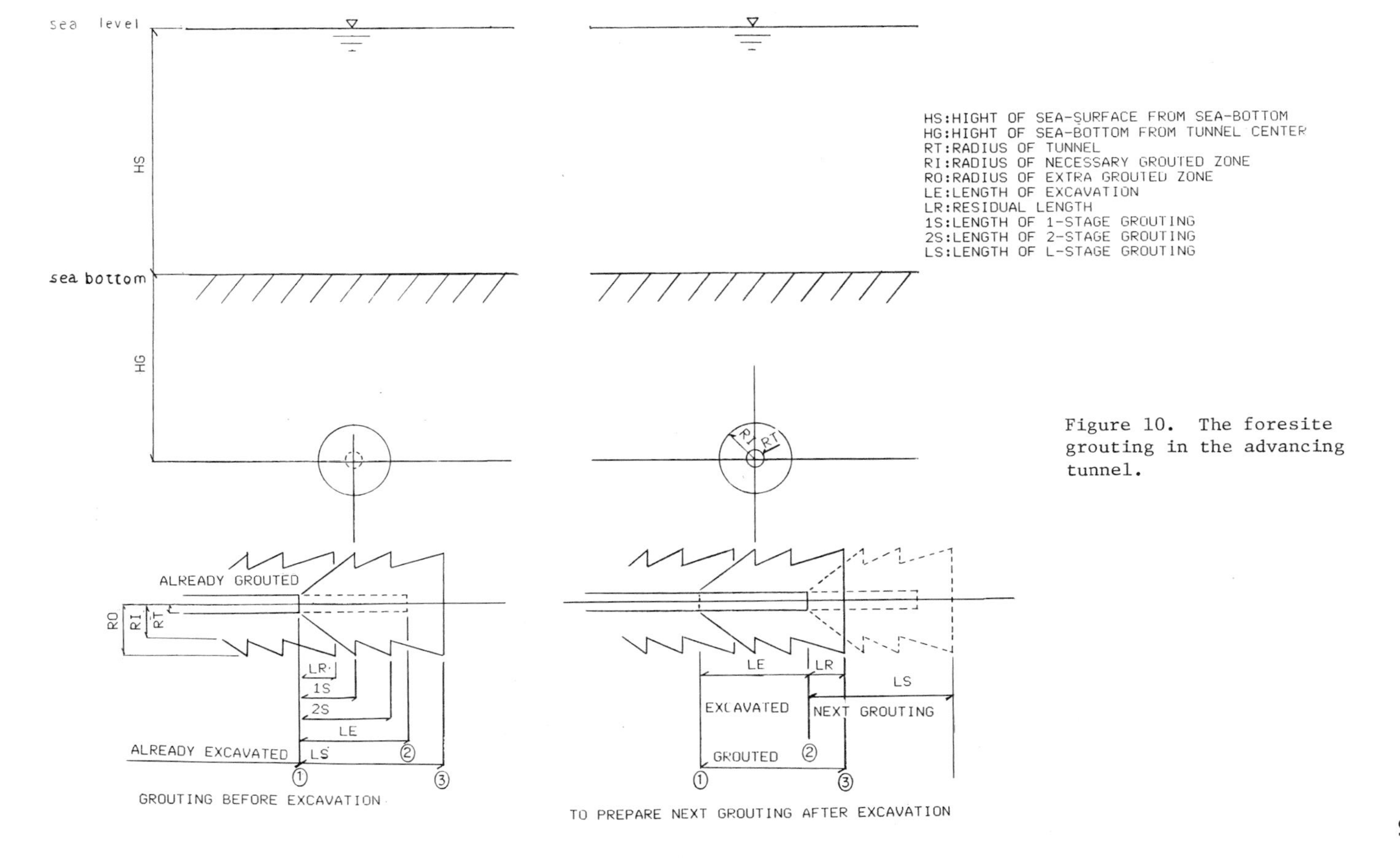

Figure 10. The foresite grouting in the advancing tunnel.

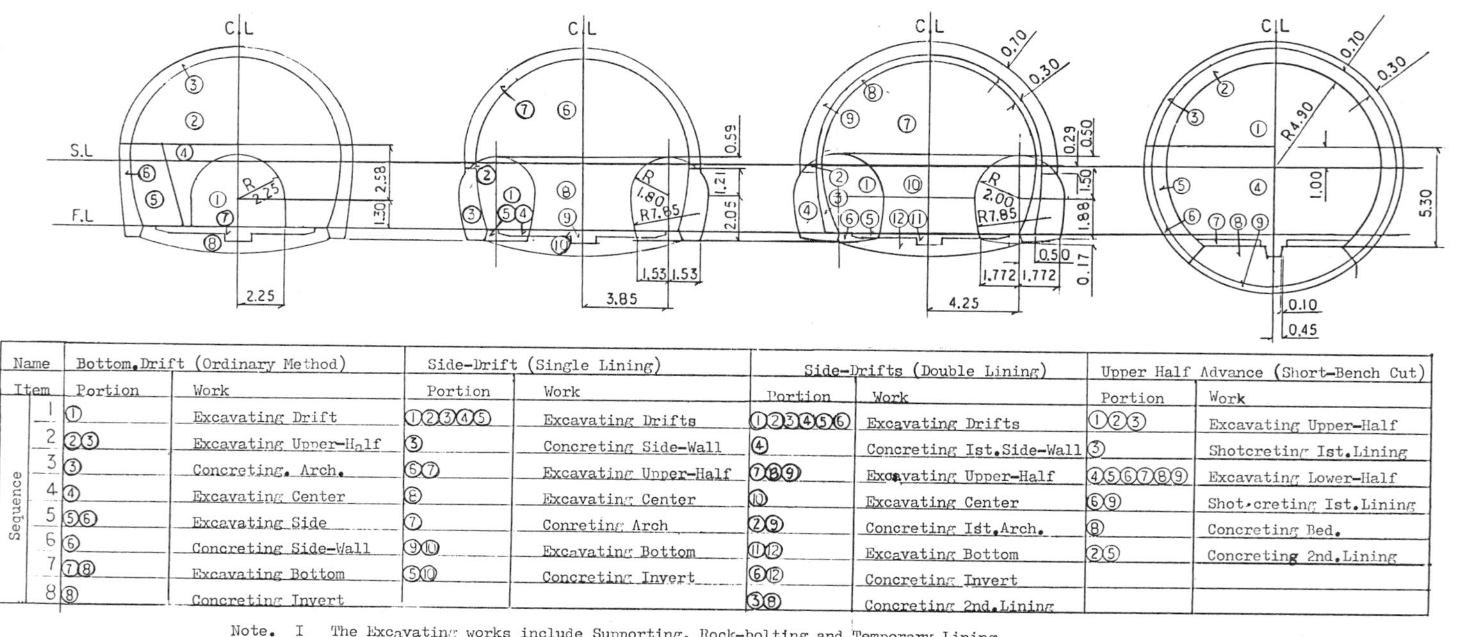

Name	Bottom. Drift (Ordinary Method)		Side-Drift (Single Lining)		Side-Drifts (Double Lining)		Upper Half Advance (Short-Bench Cut)	
Item	Portion	Work	Portion	Work	Portion	Work	Portion	Work
Sequence 1	①	Excavating Drift	①②③④⑤	Excavating Drifts	①②③④⑤⑥	Excavating Drifts	①②③	Excavating Upper-Half
2	②③	Excavating Upper-Half	③	Concreting Side-Wall	④	Concreting Ist.Side-Wall	③	Shotcreting Ist.Lining
3	③	Concreting. Arch.	⑥⑦	Excavating Upper-Half	⑦⑧⑨	Excavating Upper-Half	④⑤⑥⑦⑧⑨	Excavating Lower-Half
4	④	Excavating Center	⑧	Excavating Center	⑩	Excavating Center	⑥⑨	Shot-creting Ist.Lining
5	⑤⑥	Excavating Side	⑦	Conreting Arch	②⑨	Concreting Ist.Arch.	⑧	Concreting Bed.
6	⑥	Concreting Side-Wall	⑨⑩	Excavating Bottom	⑪⑫	Excavating Bottom	②⑤	Concreting 2nd.Lining
7	⑦⑧	Excavating Bottom	⑤⑩	Concreting Invert	⑥⑫	Concreting Invert		
8	⑧	Concreting Invert			③⑧	Concreting 2nd.Lining		

Note. I The Excavating works include Supporting, Rock-bolting and Temporary Lining.

2 The Concrete thickness include Lining by shotcreting.

Figure 11. The standards of tunneling methods in the Seikan Tunnel construction.

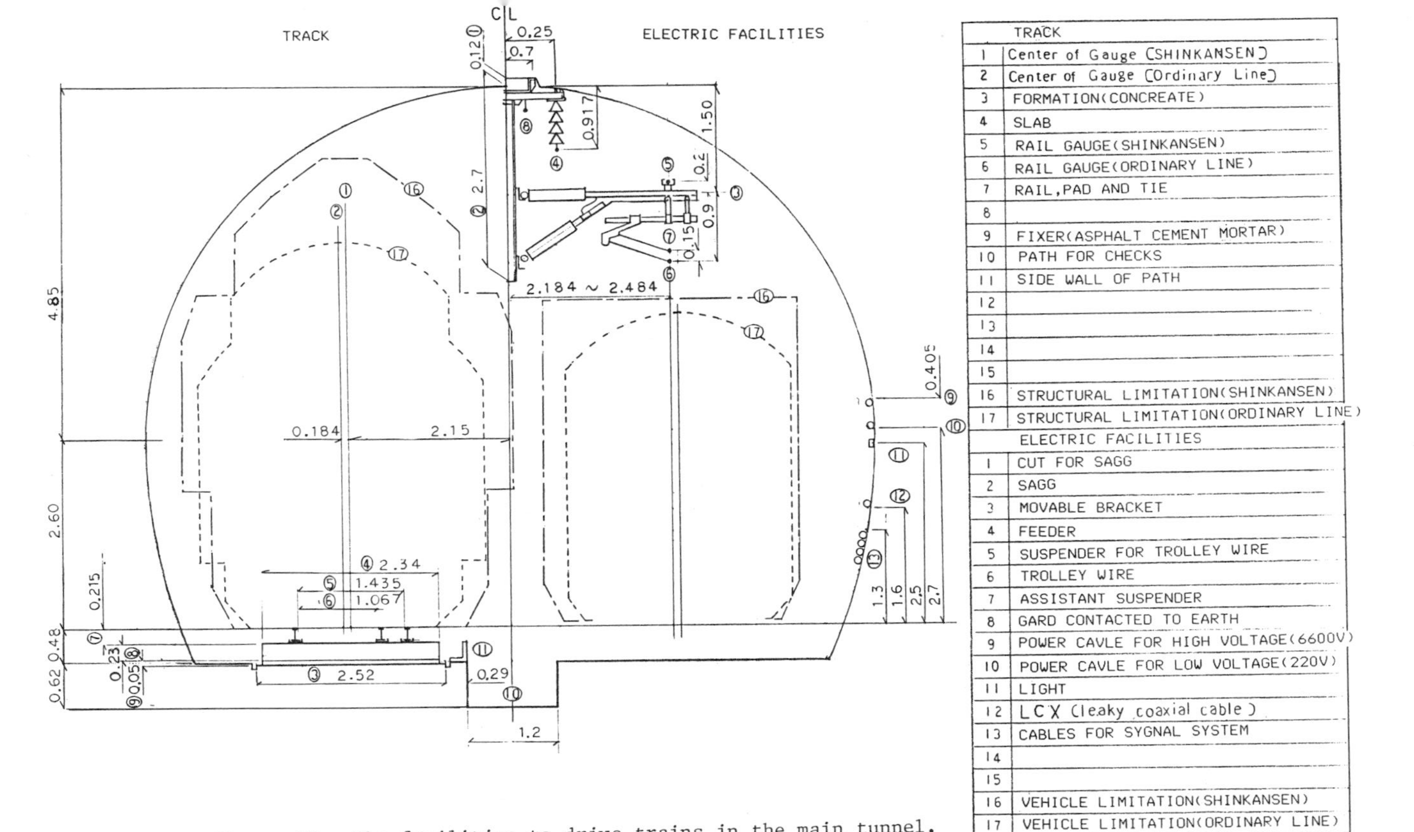

	TRACK
1	Center of Gauge (SHINKANSEN)
2	Center of Gauge (Ordinary Line)
3	FORMATION(CONCREATE)
4	SLAB
5	RAIL GAUGE(SHINKANSEN)
6	RAIL GAUGE(ORDINARY LINE)
7	RAIL,PAD AND TIE
8	
9	FIXER(ASPHALT CEMENT MORTAR)
10	PATH FOR CHECKS
11	SIDE WALL OF PATH
12	
13	
14	
15	
16	STRUCTURAL LIMITATION(SHINKANSEN)
17	STRUCTURAL LIMITATION(ORDINARY LINE)
	ELECTRIC FACILITIES
1	CUT FOR SAGG
2	SAGG
3	MOVABLE BRACKET
4	FEEDER
5	SUSPENDER FOR TROLLEY WIRE
6	TROLLEY WIRE
7	ASSISTANT SUSPENDER
8	GARD CONTACTED TO EARTH
9	POWER CAVLE FOR HIGH VOLTAGE(6600V)
10	POWER CAVLE FOR LOW VOLTAGE(220V)
11	LIGHT
12	LCX (leaky coaxial cable)
13	CABLES FOR SYGNAL SYSTEM
14	
15	
16	VEHICLE LIMITATION(SHINKANSEN)
17	VEHICLE LIMITATION(ORDINARY LINE)

Figure 12. The facilities to drive trains in the main tunnel.

Figure 12. The assisting bases to maintain the railway and countermeasures to cope with the contingency of fires on trains are constructed, besides the main tunnel under the sea. Facilities fo drainage and ventilation, and the complete equipment of the pilot service tunnels are needed to conserve the Seikan Tunnel. Approach the Seikan Tunnel which have a length of about 19 kilometers and 15 kilometers, respectively in Honshu and Hokkaido, are designed for the railway with the same specifications as the undersea portion of the Seikan Tunnel. We proceeded routinely to construct the approaches of the Seikan Tunnel. The approach tunnnels are expected to be completed in the near future, when trains will be driven through it.

The MBTA Red Line Extension

CHARLES B. STEWARD
Assistant Director of Construction
Construction Directorate
MBTA
10 Park Plaza
Boston, Massachusetts 02116

The way Boston has developed over the last century is in large measure due to the fact that we have a mass transit system, a system that's in place. The Green Line, if any of you have had an opportunity to ride on the trolley system, is the oldest subway system in the country. And for Boston, with the ocean on one side and Route 128 on the other all around the city, the decision was made a number of years ago to extend all of our rapid transit lines to the extent we can: the Green Line already goes to Route 128 on the north; the Red Line goes to Route 128 on the south, and eventually as need is felt and population densities build up, we will expand it.

What has happened, since the time when we first had our subway system and the master plans for mass transit were developed, is that we purchased the Commuter Rail system: we now have a very extensive rail system on the north side of Boston and one on the south side, and in large measure these rail systems satisfy our need to make contact with Route 128. A number of years ago when Francis Sargent was Governor of Massachusetts, the decision was made to avoid building interstate highways in the center part of the city. Plans had been made for extending Route 2 from the north down to an inner belt that circled the city, and then there would be a southwest corridor connected to I-95. Through the interstate transfer procedure, funds were made available for mass transit instead of for highways. With that funding we extended our Red Line from Harvard Square out to Alewife Parkway, and we were able to relocate the Orange Line from where it presently exists on an elevated structure in the South End of Boston onto the old railroad right-of-way which was originally taken for highway improvement. The plans that we have on the books involve extending our Blue Line up into the city of Lynn. The plan calls for an extension eventually from Alewife up into Arlington. As mentioned, the Orange Line is being relocated and in the last year we have opened our line all the way to Braintree; that was a couple of years ago, but the Quincy Adams Station which is almost at the end, was opened about a year ago. The bulk of the population shifted in Boston to the south, and that is where our need is at the moment.

Most recently, Governor Dukakis announced that we were going to open up the old Colony line, which is a railroad that extends down parallel to the Red Line and then branches out--one branch going to Brockton, one branch to Plymouth. We could look at the option of extending or reopening it for a commuter service to serve the communities on the Shore. If any of you have had the opportunity to drive on the Southeast Expressway, especially trying to come into Boston in the morning, you will appreciate the need that we have for good, efficient transportation.

Just to give credit to some of the firms that have worked on our Red Line extension: Bechtel was the overall coordinator; Harvard Square Station, Skip Marlings and Merrill; the Porter Square Station, which was the deep one, was done by Cambridge Seven; Davis Square Station by Goody and Clancy; the tunnel section from Davis to Alewife was done by Sverdrup and Parcel; and the station and garage at Alewife by Wallace Floyd.

Tunneling and Underground Transport: Future Developments in Technology, Economics, and Policy
F.P. Davidson, Editor

To help orient those who are not familiar with Harvard Square, this is the area of Harvard University. The Red Line comes down under Massachusetts Avenue. In the old days the station was right in the square, and the tail tracks and the storage and repair facilities were out in the area which has now become Kennedy Center. It was purchased originally for the library, and that did not work out, and now we have the Kennedy School at Harvard. When we started to look at the alternatives for extending the Red Line, we basically wanted to extend straight off and work our way out towards Alewife; after much community discussion, and a substantial number of community meetings and studies, the decision was to extend it faithfully up Massachusetts Avenue.

The old station was next to Harvard Yard. The new station is located at Massachusetts Avenue alongisde the Cambridge Common (where Washington assumed command of the troops during the Revolution) and then going back down to the old storage yard. This whole area is a historic district. It was a very difficult problem to try and find a way to build a massive subway system right next to two historic districts and still keep everyone happy. But the solution was eventually found.

There are bus tunnels that go under the street, under the Square, come out in this area, and permit our buses from the western suburbs to come through, drop people off in a station, and extend up to the north, reverse their run and come back. That arrangement is still in the process of being reconstructed. That is how it worked for close to 70 years.

Because of the confines of the space that we have between the stores including the Harvard Coop on one side, and Harvard Yard on the other side, the only way we could get the two functions (the bus terminals and the two subway terminals) was to offset them. And the program of reconstruction will end up doing all the paving and redecorating of the surface for everything in Harvard Yard. We have had probably a generation and a half of Harvard and M.I.T. students who have not known what Harvard Square is like, and perhaps when they come back for their graduation they will see.

As we started to build in Harvard Square, we had quite a traffic maintenance problem, protecting shoppers, pedestrians, and students while maintaining the structures in the area. And you can imagine the proximity of the excavation as we built right next to and flush with the foundation walls of some of the major buildings at Harvard.

Cambridge is an old community. In fact, one thing that is unusual about it is that it is even older than the Commonwealth of Massachusetts. So if anybody has had any dealings with old communities, I remind you that we, as an instrument of the Commonwealth, do not have any power of eminent domain over Harvard University. That happened in Lowell with the Lowell Canals. But unfortunately Harvard chose to move one of its gates back (it used to be right out on the edge of the sidewalk) in order to accommodate our construction. One of the things we found is that anything that had not been dug up by Harvard or by us or our predecessors at the subway, nonetheless had archeological value and this slowed us down at the very beginning. We discovered the complete remains of a sheep. When they first told me about it, I did not think it was terribly interesting, but on the other hand they said that in those days they would never bury a whole sheep. It either had to be somebody's pet or it died of a disease and the archeologists were trying to find it. The net effect was that we lost about a month and a half on the construction job while they dug around trying to find whatever else they could. If you have an opportunity, I would really recommend taking a look at the Red Line extension. But the area of Harvard Square down in this area is pretty much all back in shape.

The major problem has been in trying to maintain Harvard Square. We have sturdy walls right up against the face of the buildings, a lot of utility work, monstrous utility work, and we had to keep all of the stores open at all times. We tried to make everything functional and make it blend in. I think we succeeded.

One thing that we encouraged is quite an active art group in Cambridge; through early efforts, we were convinced and eventually the authorities supported us, in allocating a certain percentage of the construction funds to art. One example is a wall mural that was erected in Harvard Square. We ended up devoting one half of one percent of our construction costs to art. We thought it was going to help; we got very good support from the communities in some areas. We thought it was going to help us reduce graffiti by making people feel that the station belongs to the community.

The original plan for extending this subway from Harvard Square to Porter Square on to Davis Square was to build about 50 feet below the street up Massachusetts Avenue, an old street with old utilities. After we did the draft environmental impact statement, the results that we were getting in the meantime showed that we had pretty good rock formation down deeper, and it would be faster and more economical to move down as quickly as we could, get into the rocks, stay in the rocks, and come back up to Davis Square. So our station which started out at 50 feet, ended up at 120-130 feet at Porter; we had soft dirt down until we hit the rock in this area. This minimized an awful lot of disruption and saved us from having to build Cambridge a whole new highway. In the Porter Square area, Massachusetts Avenue comes right up this general area. We have a shopping center, we have Sears and Roebuck, we have our railroad line which goes out to the west, and in this particular case we have a station that is an intermodal transfer. Good access is provided to downtown by railroad and out to the suburbs.

As a glass cutaway model shows, you come down the escalator to the mezzanine, pay your fare, go down the long escalator into the side of the arched station. If any of you have been to Washington, it is similar to Roswell. It takes about a minute and a half just to go up the big escalator. There is a mobile hanging in the entrance, another art effort. Porter Square, for those of you familiar with it, looks like any other little group of stores on Massachusetts Avenue, and in our early community meeting, there was a strong interest by the merchants and by the community to try and put something in there that would let people know when you drove up Massachusetts Avenue that you had finally arrived at Porter Square. And out of that discussion came the answer in the form of a windmill done by a Japanese artist; it catches the wind and if the wind is too strong, they tip and dump the wind. But now at least when you drive up Massachusetts Avenue you know where you are. We also have a series of bronze gloves that cascade their way down the escalator and up in a pile in the corner.. Another art effort that the community liked.

We moved from Porter to Davis Square; this is a very dense residential area. In fact, Somerville, which is the town in which Davis Square is found, is the densest community in the United States. We own the railroad tracks that traverse this area; the subway came up onto the railroad tracks and then we were able to extend our subway on out this way.

There are other examples of art in the stations. We went to one of the local schools, the Powder House School, and the children made pictures of what they thought ought to go into the subway; these were transferred onto tiles and then the tiles were installed in the walls of the walkways. This special effort should help control graffiti.

One of the problems we have with a project of this size in a congested urban environment, building from Harvard Square to Porter to Davis and out to Alewife, is what do we do with two million cubic yards of excavated material? Fortunately, the city of Cambridge had what is referred to as the Cambridge City Dump, a landfill area which they had stopped using that they wanted to turn into a recreation area. We have a rail line that runs right next to it. As part of the program, we took excavated material out of Davis, got it into flat cars, came into Boston and came back out, and went to the city dump. We also took excavated material from here and from Harvard and the subway and took it out and deposited it in the dump. So when we were finished with this, we had a very nice recreational facility for the City of Cambridge--a very convenient way to handle all that excavated material. We ended up with our own fleet of side dump cars, which are being used by our railroad section on other projects now.

As I mentioned earlier, we made extensive use of slurry wall technology, especially down in Harvard Square. For people who are not in the construction field, we have a little diagram that also tries to show how the slurry wall cavity is first made and filled with slurry and then the reinforcing cages are installed, and how we fill it with concrete and make those panels that become the tunnel wall. This was the first major use of the slurry wall, as I understand it, in the United States, and it worked very well for us, very successfully.

Going west from Davis Square, we own the railroad right-of-way all the way into Boston, and then the Porter Square right-of-way back out in this area. There is a vital relationship of the residential area, Davis Square, and downtown Boston: the ease with which one can get to points of employment either in Cambridge or in downtown Boston is an important public convenience.

As we move further along, there are industries on one side and residences on the other. You can imagine the type of houses, a lot of wood frame houses which are typical of this area. This is where we cross Massachusetts Avenue again with our railroad line. Now we get out to where we are going to be building our new station. This area is the W. R. Grace Chemical Company property, where the line comes down and swings to make an arc so we can head up into Arlington. In this area, under the parkway, we are building a station. Going west of Davis Square, we ended up in a completely different geological situation: rock is located down at a much lower elevation; we were in very soft sand and clay all the way out.

We went across the W. R. Grace property under our new garage site and then headed up the railroad right-of-way, which we also own, out towards Arlington. As I mentioned earlier, the original plans were to have this four-lane highway, Route 2, extend on down into downtown Boston over the railroad right-of-way. We had a problem of what became known as acid sludge as we extended our subway and we found from some of the early sampling that we had pH's of 1.6, 1.7. Route 2 was designed to function with our garage; eventually when the Massachusetts Department of Public Works built ramps from Route 2, they were designed to go in, by easy access into the garage and easy access back out of the garage. And for those who are connected with Arthur D. Little, Inc., there will be a sidewalk that goes right alongside so they can get to the subway very easily. It is a high-capacity garage. We designed it to have 12 berths for buses. It is a 2,000 car garage, and notice that three-quarters of its capacity is scheduled to be filled in the rush hour. There is a number of spaces for short-term parking; we expect a high number of walk-in arrivals, and we have provided 100 spaces for bicycles. We have a very active bicycle group in this area, and we have provided space here and at our other stations for bicycle riders.

We had problems as we built in this area; fortunately, we had very active community participation. There were meetings every two weeks. All the communities that were in that area were involved, all the neighborhoods were involved, Sierra Club was involved. In fact, the Sierra Club supported us strongly for what we were trying to do. There were people who were concerned about the wetlands here. We were in the flood plain, so we had a lot of problems dealing with that. A very active organized group opposed what we were trying to do and tried to oppose construction of the Red Line completely, under the guise of not hurting the river under which we passed via a tunnel.

Using a light well in the middle, we tried to make the mass of the garage not so oppressive. We also found that, if we took the budget and we had allocations for benches in the very beginning, instead of simply buying benches off the shelf we went to craftspeople and said, "Would you give us X number of benches for this money?" we got excellent value for the same price and the benches are very well liked.

Upstairs on the ground level of the parking garage we have a glass-enclosed waiting area for people who are waiting for the buses. In fact, we even tried to provide heating with some heat scavenged out of the subway.

If there are any questions, I will be happy to try and answer them.

Question: You made reference to the decision to go up Massachusetts Avenue as opposed to a straight line, without indicating why that decision was made.

Answer: The shortest distance and easiest would have been what was called the Garden Street Alignment, which was simply to go from Harvard Square straight out to Alewife, but there were strong concerns from the city of Somerville and Cambridge residents, some segments of the Cambridge community, who wanted service to come to Porter Square and to provide service in Davis Square. Somerville is one of the inner cities that serves a heavy concentration of bus lines by the MBTA, and they felt as one of the inner cities and the only inner city not having a subway, that they would like one, and out of that long dialogue over a period of time came the decision to go up Massachusetts Avenue to Porter, then over to Davis, then out to Alewife.

Question: What was the total cost of the project?

Answer: On the order of about $580 million.

Question: Did you ever calculate what that sheep cost?

Answer: Not really. We lost about a month, but the contractor was able to work in other areas. We have a very good relationship, I would say, with the Massachusetts Historical Commission. They were very cooperative in trying to expedite the process any time we ran into a conflict.

Question: Cambridge is probably the most environmentally sensitive community in this area, as Arthur D. Little has discovered. When you found the acid sludge, what were you able to do with it?

Answer: In the very beginning, when we realized we had a problem, we got in touch with our Department of Environmental Quality, the EQE people. Eventually, the conclusion was that we would go through a solidification process patented by a firm out of Macon, Georgia, outside of Atlanta. It consisted basically of mixing the sludge material with their process, and what resulted was a material that looked like crumbly concrete which when tested satisfied the environmental people and was used as a cap in a number of landfills in the area.

Question: Were any federal actions involved that accelerated or decelerated the process?

Answer: Well, we have very good support from our local UMPTA office. That helped a lot in expediting the process. There were two major things, I think, that complicated the process. I will just go back and say as a policy we found the best way to solve a problem is doing it the right way. We tried identifying every problem, every community group, and every environmental issue as early in the process as we could and take care of issues promptly so they do not show up later and cause us unnecessary vexation.

There were two things that came up over the years. One was what was called a peer review process, and unfortunately I think from our point of view ended up being towards the end of the design phase. We wanted to get into construction, and the Washington office of UMPTA wanted us to go through this process of reanalyzing a lot of the decisions that had been made. This caused, I would say, about a 12-month delay in our ability to build a tail end of the project. The original idea was to have Alewife and Harvard open at the same time, and as a number of you know, Harvard has been operating for some time and Alewife just opened up a couple of weeks ago. I have nothing against the peer review process. In fact, it is very good and we are using it on a lot of our other projects. But it should occur early in the design stage, not at the end when it has a tendency to cost money. That we figured was worth about 12 months. In an overlapping fashion it was about another 12-month loss when we tried to get our permit from the Corps of Engineers to build under the Alewife Brook. And the combination of those two counted for about an 18-month delay in our ability to have the whole line open at the same time.

Question: The town of Arlington continues to oppose the extension. That resolves the issue, does it not?

Answer: Which issue?

Question: Extending the Red Line.

Answer: Our master plan still calls for extending it up through Arlington through Lexington to Route 128. And I would say that at some point the demand will be enough to justify building the line to that point.

Question: Will the demand be enough to create the environment locally for it? Are you suggesting that the state will decide to do it irrespective of the sentiments of Arlington?

Answer: Well, Arlington is concerned, as I understand it. There were social concerns that came out in a lot of our meetings that if you build a subway you will have a change in the makeup in the community. I think over a period of time that will go by. They also did not want to be the end of the line. That is what we refer to as "terminal syndrome": we do not understand why it is bad to be the end of the line because we have had a lot of such places that are very happy. The promise that was made was that if we build into Arlington we will build all the way to Lexington and it would not be the end of the line. It would make sense if we continue to really build to Arlington Heights on the Lexington line, and if we can build further. But I do not see the demand for that from a transit point of view, certainly for another 20-30 years.

We have very limited, or what we consider limited, resources, and I would say there are other areas that people are really anxious to have us build, for instance, the governor's message that we should go back to the South Shore with the commuter rail. And I think our attention will be spent in areas like that.

Question: North Station as well?

Answer: North Station. Central subway and the old parts of it.

Question: One final question. Have you ever analyzed the impact of the Red Line on real estate in the affected communities, and can you estimate what Arthur D. Little, Inc., property is worth today as opposed to pre-project?

Answer: On the second one, I do not really know the answer to that one. Early in the environmental process we had a number of market studies that were done that appeared to demonstrate that if we built the Red Line it would be positive for the economies of the area. As we have been building the subway for the last seven years, I, having lived in this area for some time, have been very impressed with the amount of private investment that has gone on in Harvard Square, up Massachusetts Avenue, Porter Square, and Davis Square.

New High-Speed Lines of the German Federal Railway

WILHELM LINKERHAGNER
Head of the Department of New Railroad Lines
Frankfurt/Main, West Germany

INTRODUCTION

In 1985, railways in Germany reached 150 years of age. On the 7th of December, 1835, the first train--the legendary Eagle--traveled from Nuremberg to Furth, watched by many incredulous spectators.

The invention of the railway was the precondition for the rapid industrial development having taken place in Europe and North America during the 19th century. Very shortly it had become number one among all transport operators with monopoly character. This position could be held for almost 100 years.

The railway was already losing its supremacy in traffic after World War I, but to an even higher degree after World War II. Today, it competes heavily with other transport operators, such as roads, waterways, air navigation, pipelines, and high-voltage lines. If the railway wants to overcome and to survive this competition, it has to improve its scope of services decisively.

BASICS FOR PLANNING NEW RAILWAY LINES

Transport Route Planning

The last trunk lines within the network of the German Federal Railway were built about 100 years ago. Many aspects of this network do not meet today's requirements for a high-grade demand in passenger and freight traffic:

-- The main traffic flows in Germany have changed from east-west to north-south direction.

-- In many cases, small radii and large gradients are reducing the admissible maximum speed to less than 62 miles per hour.

-- The main routes are overstressed, whereas the remaining network disposes of underutilized capacity.

In order to increase its competitiveness, the German Federal Railways are planning on expanding a core network being about 2,000 kilometers long, on which maximum speeds of up to 156 miles per hour can be achieved. This core network shall connect the centres of main settlement and economy in the Federal Republic of Germany. Passenger trains having a travelling speed of up to 112 miles per hour and freight trains with a travelling speed of up to 62 miles per hour are to make use of this network.

The route planning of the German Federal Railway has been integrated into the Federal Transport Route Plan. This plan is issued by the Federal Government for all Federal transport routes--roads, waterways, railway lines, and airports. Only measures of which use in national economy is to be expected will be integrated into the Federal Transport Route plan. This plan is revised every 5 years.

Tunneling and Underground Transport: Future Developments in Technology, Economics, and Policy
F.P. Davidson, Editor

The announcements of the German Federal Railway for the Federal Transport Route plan are comprising the new construction and upgrading of routes having a total length of about 2,000 kilometers. The whole program has an investment of approximately $12 billion. Six billion dollars of this amount has already been granted by the latest revision of the Federal Transport Route plan in 1980. The pending sum of $6 billion for route expansion has been applied for by the German Federal Railway in order to dispose of a network being optimal as it concerns operation and national economy, so that it can be used for maximum-speed railway traffic. The German Federal Railway aims at investing about $0.7 billion per year for the improvement of the route network during the coming 15 years.

The plannings of the German Federal Railway are part of the infrastructure master plan for the European railways. This plan has been drawn up by the International Railway Union (UIC). The new construction and the upgrading of important routes of long-distance traffic on rails is the principal aim of the European Infrastructure Master plan. Because of the central location of the Federal Republic of Germany, the route network of the German Federal Railway is a core part of this traffic conception borne commonly by both east and west. The plan itself determines standards for a highly efficient rail network of about 30,000 kilometers length.

ECONOMY

Since the last war, the German Federal Railway's participation in traffic has constantly decreased. Its percentage in freight traffic only amounts to 30% and in passenger traffic to only 7%. For ecological and economic reasons, it is a definite political declaratory act on the part of the German Federal Government that the railway should make an essential contribution to total traffic also in the future. A special advantage of the railway is to be seen in heavy-flow passenger and freight traffic as well as in public transit in metropolitan areas. Therefore, the main focus of infrastructure investment is concentrated in these areas.

When drawing up the Federal Transport Route Plan, the infrastructural measures announced by the various traffic operators are subject to a cost-benefit analysis with respect to national economy. The measures of the German Federal Railway are also examined in view of business administration. At present, two new railway lines, from Hannover to Wurzburg, and from Mannheim to Stuttgart, with a total length of 426 kilometers are under construction. The cost-benefit relation concerning the national economy of both lines amounts to about 4:1. Thus, the benefit from the two lines is about four times higher than the investment and derivative costs. After completion in 1991, the economic result of the German Federal Railway will annually improve by $200 million; in this way, the measure also makes sense as it concerns business administration. The new railway line is financed by the Federal Government. For the examinations in national economy and in business administration a real interest rate of 3.5% will be calculated.

Effects on the Environment

In Central Europe, the density of population is extremely high. For this reason, environmental compatibility is becoming more and more important when constructing new traffic routes. Here the railway keeps a superior position among all traffic operators.

-- Energy Consumption and Air Pollution (Fig. 1). In passenger traffic, the same traffic power requires three times more energy when using the car, and even five times more when using the airplane. In freight traffic, the relation amounts to 1.3 for inland navigation and to 3

Energy Consumption
Passenger Traffic

1 IC Train
3 Car
5.2 Airbus

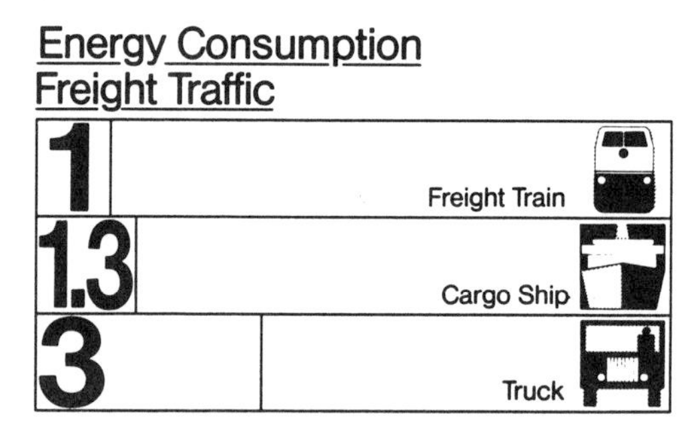

Figure 1. The energy consumption of the different traffic vehicles.

for trucks. This shows that the railway causes the least air pollution, all the more that the most important part of the power is gained by electrical traction being harmless to the environment.

-- Ground Consumption (Fig. 2). Also under this aspect, the railway keeps a leading position among all competing traffic operators. In order to achieve standard traffic power, the railway needs about one third of the space of a motorway and about one fourth of the space of a waterway for inland navigation. This criterion is of major importance in densely populated Europe.

-- Safety. One injured person in railway traffic compares with 106 injured persons in street traffic (at the same conveying power). The relation amounts to 1:26 as it concerns dead persons. That means that the railway is much safer than any other transport vehicle.

Design Parameters

The Federal Republic of Germany is showing a decentralized structure in population and economics. The traffic flows are correspondingly differentiated and numerous. Therefore, it does not make much sense to build traffic routes solely used for passenger or freight traffic. With a maximum speed of 156 miles per hour, the new railway lines of the German Federal Railway serve for rapid conveyance of passengers on long distances as well as for the highly graded freight traffic with a peak speed of 75 miles per hour. The design parameters are determined by the above-mentioned conditions and by physical limiting values, comfort characteristics, maintenance expenditure, and by the amount of the first investments (Fig. 3).

Ground Consumption

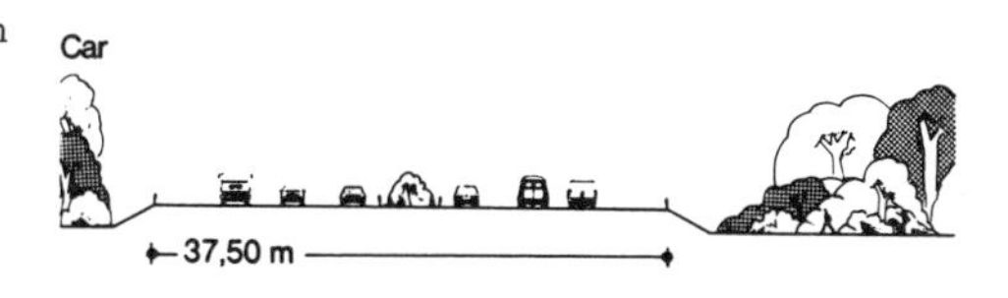

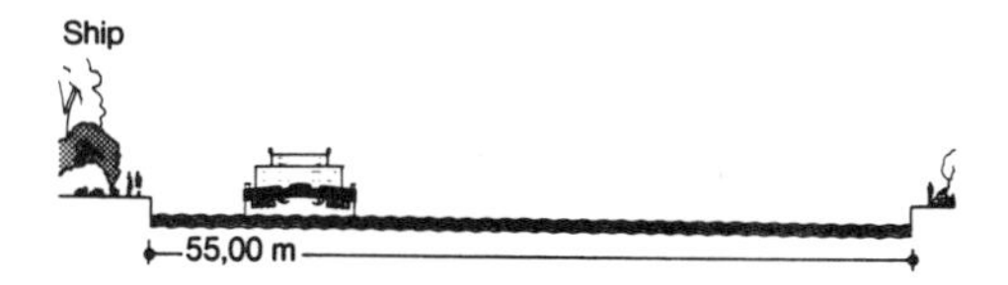

Figure 2. The ground consumption of different traffic vehicles.

Figure 3. The design parameters of the new railroad lines.

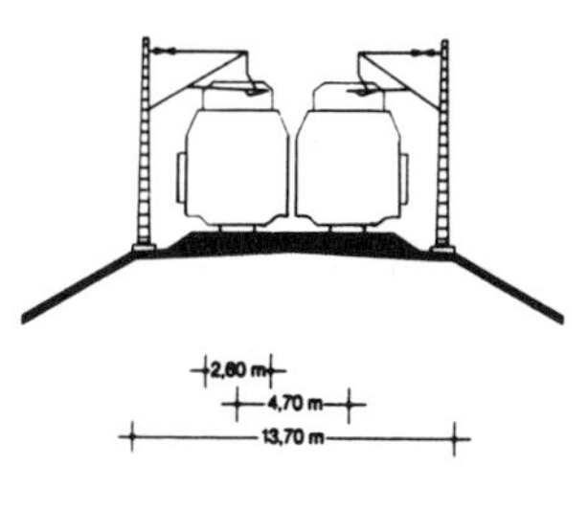

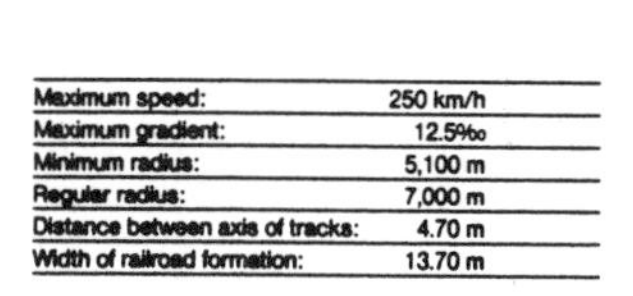

Maximum speed:	250 km/h
Maximum gradient:	12.5‰
Minimum radius:	5,100 m
Regular radius:	7,000 m
Distance between axis of tracks:	4.70 m
Width of railroad formation:	13.70 m

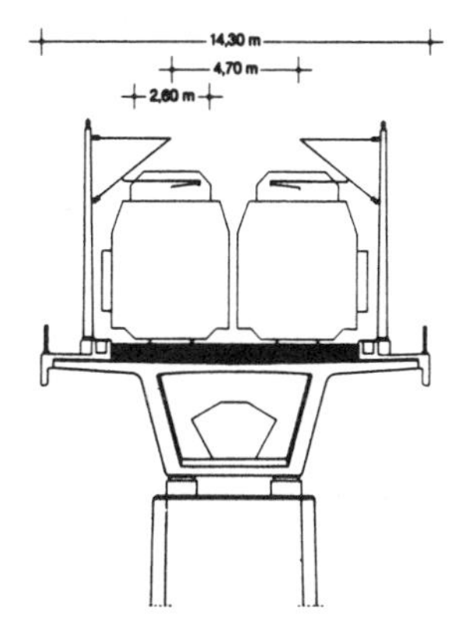

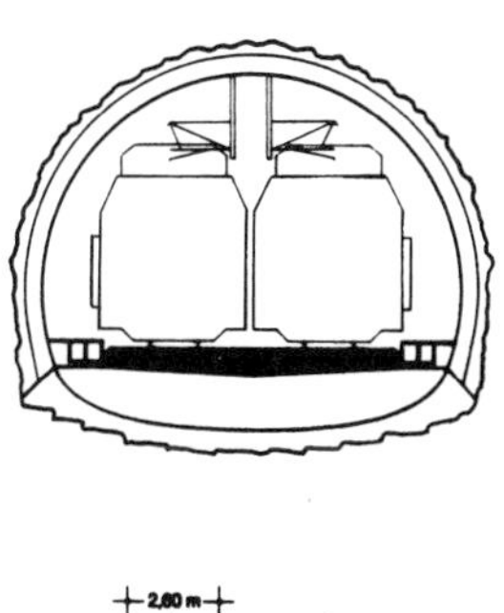

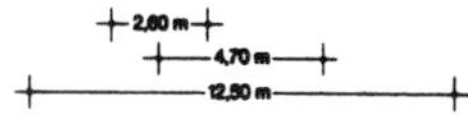

The largest longitudinal gradient amounts to 12.5, which means that freight trains with a total weight of 1,200 tons can be driven with one engine. The smallest arch radius amounts to 5,100 meters. For reasons of maintenance 7,000 meters should usually not to be fallen short. The distance between axis of tracks amounts to 4.70 meters, and the width of the railroad formation to 13.70 meters.

The carriageway will be formed as ballasted track. The rails weigh 60 kg/m, the concrete ties are 2.6 meters long, they weigh 300 kilograms and will be placed at a distance of 0.6 meters.

The routes are electrically run by two-phase alternating current of 16 2/3 Hz and 15 kV. The power supply is effected by its own traction current conduction of 110 kV.

Execution of Construction

At present, both new railway lines, from Hannover to Wurzburg and from Mannheim to Stuttgart, are under construction. The costs run up to $3.7 billion for the Hannover-Wurzburg line and $1.2 billion for the Mannheim-Stuttgart line (price level 1984). Construction was started in 1980. The current schedules earmark the completion of both new railway lines for 1991.

For the most part, the new railway lines pass through the hilly highlands. Because of the low gradient and the large arch radius, the new lines have numerous tunnels, bridges, and earth structures. 36% of the new Hannover-Wurzburg railway line passes through tunnels, 11% on bridges, 28% in cuts, and 25% on embankments.

TUNNEL CONSTRUCTION AT THE NEW RAILWAY LINES

General

One hundred fifty kilometers of the new railway lines pass through tunnels. The tunnel structures represent the core part of the construction measures. All tunnels are driven by the New Austrian Tunneling Method (NATM) and have a double-shelled lining. The outer shell usually consists of steel arches, wire meshes, shotcrete, and anchors.

If necessary, the inner shell is supported by a waterproofing membrane or built with watertight concrete. In case the rock conditions are difficult, an arched concrete bottom, afterwards filled with lean mixed concrete, is applied.

Size and form of the excavated cross section depend on the following marginal conditions:
-- railway clearance
-- thickness of expansion
-- construction method.

From the given distance between axis of tracks of 3.70 meters follows a required effective cross section of 82 square meters on the straight line. This effective cross section is 40% larger than the one of conventional two-track railway tunnels; this has an extraordinarily positive effect on the aerodynamic conditions when two trains meet each other in the tunnel. On straight lines, the required excavated section amounts to 105 square meters; in arched secondary tensioning areas 145 square meters of cross section are to be excavated.

In April 1985, about 50 kilometers--one-third of the total tunnel length--was driven. Thirty tunnels are under construction, in the course of which the work at more than 40 working faces is executed at the same time. Those are enormous achievements, which are momentarily made by the German tunnel construction industry. On average, 1 kilometer of the framework costs about $10 million; this corresponds to a price per cubic meter of $90. When comparing these prices, it nevertheless has to be considered that the momentary rate of $1 = 3 Deutsch Marks does not allow a realistic comparison, measured at the living standard.

Geology

The tunnels of both new railway lines are mostly located in the triassic. New red sandstone, limestone, and keuper belong to this geological formation. The rock characteristics are reaching from hard stable rock to completely softened and loosened rock. The tunnels are situated partly above and partly below the ground water level.

Extensive and systematic geological preliminary investigations were made before the construction measures had started in order to offer in the tender documents a serious calculation basis for the submitting construction companies. For the 327 kilometer long new railway line from Hannover-Wurzburg, 2000 core drillings with a total length of 80 kilometer of drill cores have been made.

These drillings showed that the geology is strongly varying and that totally different rock characteristics may occur even within one tunnel.

The so-called syncline pipes in the middle new red sandstone are especially difficult. In geological periods, leeching zones have formed in the depth of the earth crust; the zones have broken through in the course of years. With time, these tubular cavities (diameter 10-30 m) have become filled with a clay-like material. Such remolding zones may occur unexpectedly when driving the tunnel; so they require an immediate adaptation of the construction method to the difficult rock conditions.

The presence of anhydrite being capable of swelling is characteristic for the keuper formation of the new railway line from Mannheim-Stuttgart. If water is entering, the anhydrite is changing into gypsum. This metamorphosis leads to an increase in volume by about 60%. As a result, the dimensions of the tunnel lining for these swelling pressures is 2 meters thick and strongly reinforced concrete shell. In order to avoid this cost-intensive expansion, the entry of water has to be completely stopped during the drive.

The 6.6 kilometer long Freudenstein Tunnel traverses the gypseous marl in several intersections at very small angles. For the exploration of the rock characteristics a pilot tunnel (diameter 4 m) is presently driven (Fig. 4). In the course of the construction works of this pilot tunnel, extensive theoretical and practical investigations are carried through in order to gain knowledge about realization of the economical drive of the main tunnel. The excavation of a test chamber belongs to this exploration program. In this chamber, irrigation experiments on scale 1:1 will be executed in order to analyze the stress and deformation behavior at different sizes and forms of expansion. The costs of the test tunnel including the investigation program, will amount to approximately $30 million.

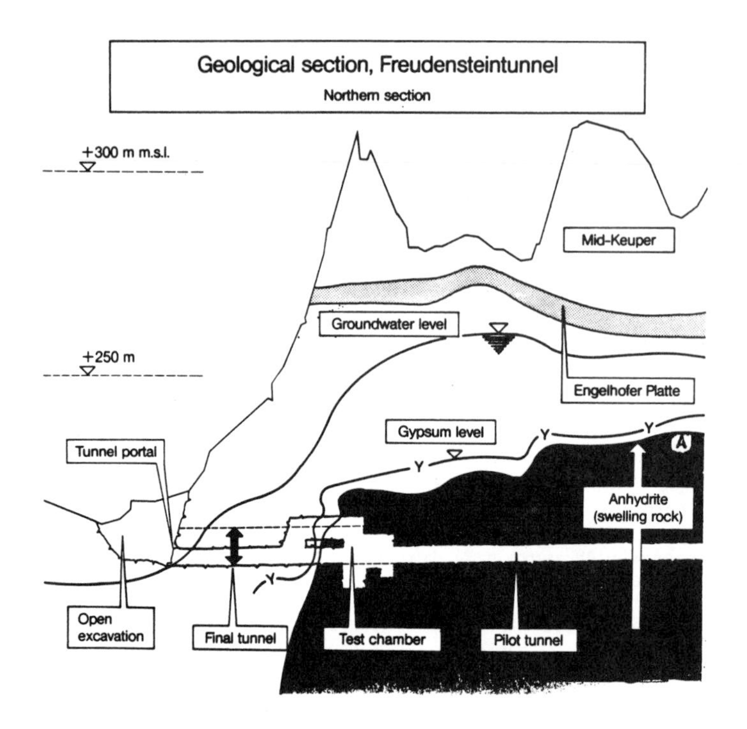

Figure 4. The Freudenstein pilot tunnel in swelling rock.

Choice of Tunneling Method

In the course of the new railway lines about 70 tunnels with a total length of 150 kilometers will be driven. Before the start of the construction measures, the German construction industry and the German Federal Railway carried out profound examinations about which method should be applied for tunneling. The use of full-section machines and the NATM were discussed.

The result of detailed examinations[1] was that in consideration of the present marginal conditions of geology, contract and labor regulations, the NATM was most suited. Even tunnels of major importance (two tunnels with more than 10 kilometers in length) showed that the use of full-section machines is uneconomical. In the following, some decision criteria are explained.

Construction Time. As exemplified by the 10 kilometer long tunnel, the difference in construction time between the NATM and the full-section machine had been examined (Fig. 5).

-- Development and construction of a full-section machine took about 15 months.

-- The operating power of a full-section machine is relatively high, but it can only work one sided.

-- The preparation time of the NATM is very low; directly after the award the construction can be started.

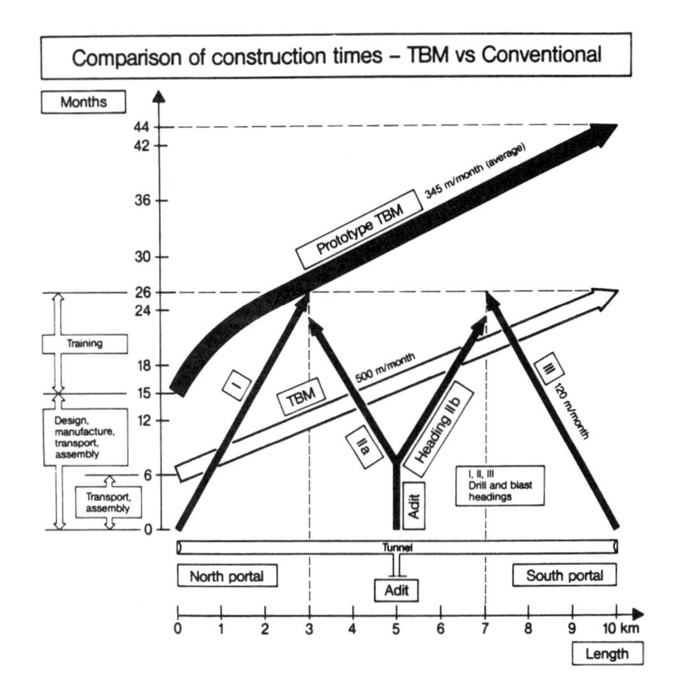

Figure 5. The construction time of TBM and NATM.

-- By means of the NATM sometimes 4 faces can be worked at the same time, because of the mostly shallow cover, which allows the drive of a lateral access tunnel.

-- So, the NATM can be applied quicker, even at long tunnels. Since all tunnels of major importance are located on the critical way, this is an essential criterion of decision.

Cross-Sectional Area. Cross-sectional areas of up to 145 square meters are not yet being driven with full-section machines until now. Through primarily reflections, the following knowledge has been gained:

-- Full-section machines cut a circular profile toward the required tunnel with a core profile; this leads to an extra excavation on the order of 10-20%.

-- The development of elliptical profiles gets under way; until now an economic advantage has not yet been seen.

-- The large excavated cross sections require the raising of strong driving forces. The introduction of these forces into the rock is partially problematic.

Costs and Risks. There are different cost structures in Europe and the United States:

-- The equipment for a tunnel being driven by the NATM from two sides amounts to approximately $3 million. The purchase of a full-section machine requires investment of about $12 million.

-- When driving with a full-section machine, high machine costs and relatively low costs for wages will occur, whereas when applying the NATM, it is vice versa. Europe has lower wages than the U.S. Furthermore, Europe disposes of a sufficient number of engineers with experience in NATM.

-- For reasons of the remolding zones often occurring unexpectedly in the highlands, there is a high risk of a total breakdown when using a full-section machine.

Profound examination before starting with the construction measures has shown that in the German highlands, the NATM is more economical, because of its higher flexibility, its favorable construction time, and its lower risk of a total breakdown.

New Austrian Tunneling Method (NATM)

The NATM is known all over the world. In Central Europe it is the standard tunneling method, even for the most difficult rock conditions. The essential principles are listed below:[2]

Bearing Effect of Rock

-- The essential bearing component of a tunnel is the rock itself.

-- In order to accomplish this function, the rock must be treated smoothly.

-- By means of the timely placement of support the softening of the rock will be prevented.

-- Stress concentrations destroy the rocks; they can be avoided by rounded cross-sectional forms.

Support

-- The support must be placed at the right moment.

-- The deformations required for arch formation have to be allowed for.

-- Over-sized deformations leading to softening have to be prevented.

-- A flexible support will be achieved by shotcrete.

-- In difficult rock the ring closure has to be established rapidly in order to stabilize the rock.

Control and Supervision

-- Deformations and stresses are continuously measured.

-- The dimensioning is steadily fed back from the measuring results.

-- The means of support can be adapted to changed rock conditions at any time.

-- Critical construction conditions are announced early, so that support measures can be taken.

If all these principles are observed, a tunnel can be driven economically and safely with the NATM, even in the most difficult rock. As it concerns the tunnels of the new railway lines, the following procedures have turned out to be efficient (Figs. 6, 7):

- On account of the large cross-sectional area the tunnel will be driven in 3 steps--calotte, bench, bottom.
- Depending on their solidity, the rocks will be loosened by smooth blasting, roadheaders or excavators, and removed by trucks.
- Directly after the loosening process the excavation soffit will be sealed by shotcrete in order to prevent the near-surface softening.
- Afterwards, steel arches will be put up and wire meshes will be fixed. Then, the final shotcrete layer will be applied.
- For the increase of the bearing effect 4-to-6-meter-long cement mortar anchors will be placed.
- The subsidence of the bench and the placement of the bottom are carried out afterwards in a distance of 20-150 meters, depending on the rock conditions.
- Independent of the actual drive the final lining is placed by means of a formwork.
- All work processes are controlled and supervised by means of continuous deformation and stress measurements.

Innovations in Tunnel Construction

Up to now, the NATM has proved to be good at the holing-through of the German highlands. Because of the hard competition, presently dominating the German tunneling market, various innovations have been promoted which had participated in the further development of tunneling technology.

Shotcrete

Usually the shotcrete is applied by means of the dry spraying method, where the concrete aggregates, cement and water, are transported separately to the nozzle. When reaching the nozzle, water will be added to the dry cement/concrete aggregate mixture. Since the regulation of water addition is difficult, a relatively high bounce will occur, which leads to high material costs. Furthermore, the working conditions for the nozzle leader are bad because the development of dust is very high when using the wet spraying method.

At several tunnels of the new railway lines, the shotcrete is applied by means of the so-called dry spraying method, where the concrete is pumped to the nozzle as a finished mixture. Because of the better possibility for proportioning of the concrete aggregates, the bounce is low (about 10%). Since the development of dust is also low, there is only little sanitary strain for the nozzle leader, all the more that the Colgunite is no more applied by hand-guided nozzles, but with spraying machines (Fig. 8). The major advantage of the wet spraying method is the high early strength, which brings high driving power, short

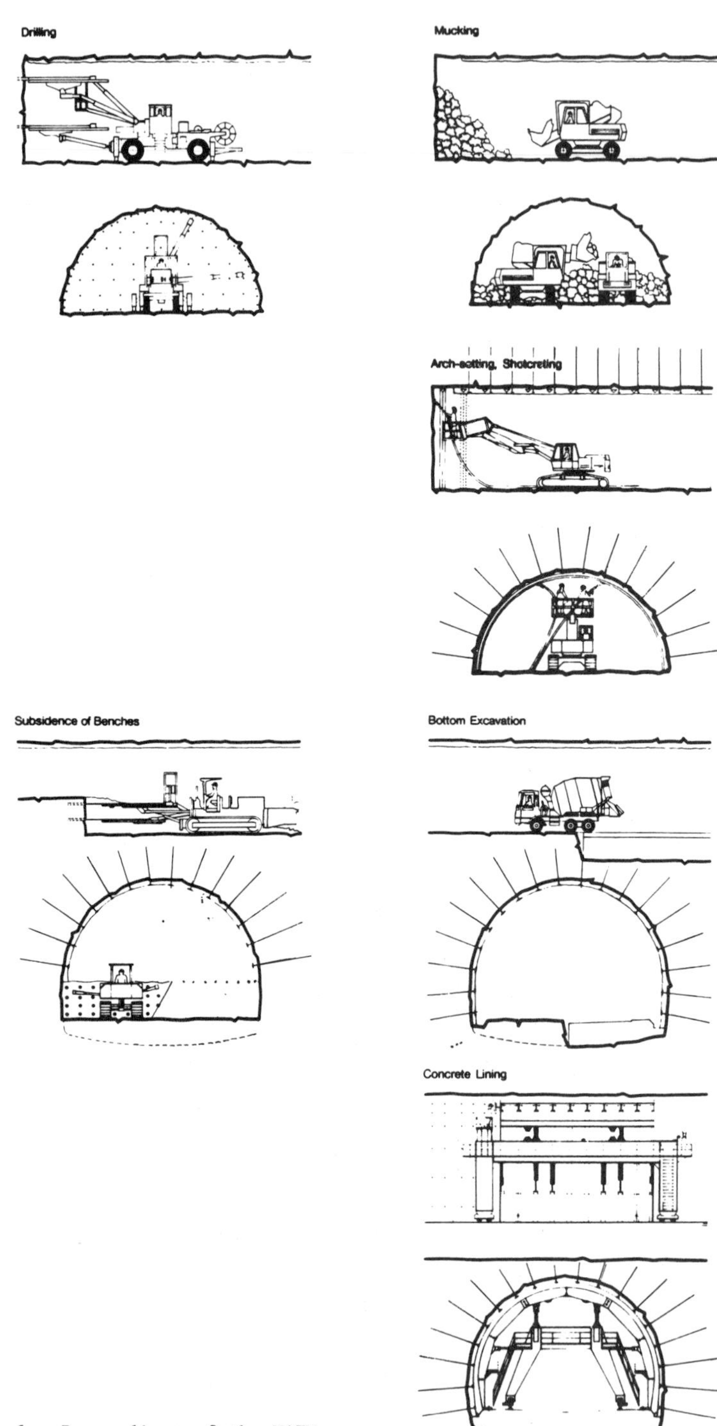

Figure 6. Proceedings of the NATM.

Figure 7. The Rollenberg Tunnel; Calotte, bench and the pilot tunnel at the bottom.

Figure 8. Using the wet spraying method, the shotcrete is applied by spraying machines.

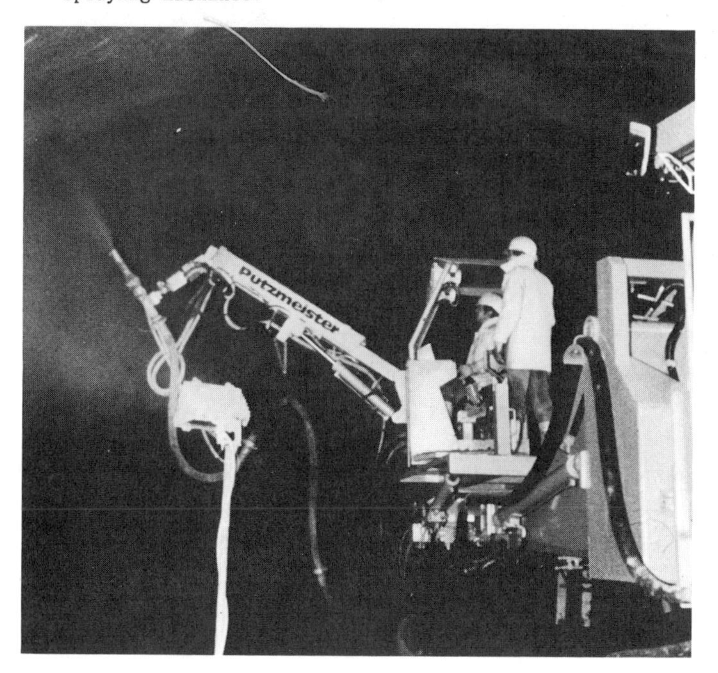

construction periods, and therefore, low costs. The only critical aspect of this new method still is to make sure that the quality of execution is equally good.

Steel Arches

Until now, in tunnel construction steel arches had been placed as solid profile. Grid arches are gaining more and more importance. They are lighter, and therefore more favorable in price than comparable solid profiles; they are easier to handle and can be injected better because of the profile with several drifts.

Final Lining

At tunnels with entering water, the final lining will either be sealed by a waterproofing membrane or constructed with watertight concrete. At tunnels being sealed by a waterproofing membrane there usually is one problem that means that the reinforcement suspended at the fixing points of the membrane might easily cause damages at the membrane. For this reason there had been erected for the first time a self-contained reinforcement, which has no contact with the membrane. So damages at the membrane occurring when placing the reinforcement can largely be avoided (Fig. 9).

Figure 9. Sealing the tunnel by a waterproof membrane and placing the reinforcement.

The development of watertight concrete had been advanced during the last few years. The problem is the limitation of the crack width, which partially leads to a high reinforcement percentage, and so to high costs. However, the concrete technology has proceeded so far, that today watertight concrete is used for many tunnels of the new railway lines.

Conclusion

Like hardly any other construction method, the NATM requires a close cooperation between the client, the designer, and the construction company. There are constantly emerging decisions of how to continue the work; they cannot be arranged behind a desk, but exclusively on the spot. It is extraordinarily important for the quality of the structure that all participants have sufficient practical experience in tunnel construction apart from profound theoretical knowledge. The multitude of the drives taking place at the same time in Germany has the consequence that the capacities of appropriate personnel for design and building sites are momentarily nearly exhausted.

CLOSE

At the present time, railways are constructed all over the world. This shows that also in the future, major importance will be awarded to the rail traffic. The German Federal Railway is assuming that the railway's attractivity soon will increase because of its travelling comfort, its harmlessness toward the environment, and its low specific energy consumption. In order to be able to offer favorable traffic services in future, German industry and the German Federal Railways are commonly developing a maximum speed train, the ICE (Fig. 10). In March 1985, the prototype was introduced to the public. In 1991, the client will be able to travel rapidly, punctually, and comfortably through Germany at 156 miles per hour.

Figure 10. ICE, the experimental train for high-speed railway traffic in Germany.

LITERATURE

1. Distelmeier/Semprich, 1985, New Methods of Constructing Railway Tunnels in Germany, Tunnelling, Brighton, Great Britain.

2. Muiler-Salzburg/Fecker, 1978, Grundgedanken und Grundsatze der "Neuen Osterreichischen Tunnelbauweise," Felsmechanik Kolloquium Karlsruhe.

;unken Tube Tunnels Construction: Typical Projects—Past ınd Present

IARTIN N. KELLEY
resident
iewit Engineering Company
400 Kiewit Plaza
maha, Nebraska 68131

. HISTORY OF TUBE PLACEMENT

The first sunken tube project was a railroad crossing of the Detroit .iver between Michigan and Ontario, Canada, completed in 1910. This tube 'as approximately 2,500 feet long and was built to carry two lanes of ailroad tracks. The first sunken tube highway crossing was also done in his country and was the Posey Tube in California, completed in 1928. It 'as a two-lane crossing, approximately 2,400 feet long. It was actually he third sunken tube project built in this country, but it was the first ighway crossing. From the time of the first tube in 1910 until 1950, a eriod of forty years, six tubes were constructed in North America. Six ore sunken tubes were completed between 1950 and 1960. Five were ompleted between 1960 to 1970, and six were completed from 1970 to 1980. he Fort McHenry Tube and the second Downtown Elizabeth River Tube are the 'irst two to be completed in the 1980s. For a listing of all the projects ince 1950, see Table 1. Of the 19 tube projects listed in the table, 3 re cast-in-place concrete and 16 are steel shell and concrete lined. The ubes have been for both highway crossings and railroad crossings. The ighway tubes have been up to six lanes wide, in the case of the oucherville Tube in Quebec, Canada, and they have been up to four ailroad or rapid transit tracks, in the case of the 63rd Street Project n New York City. The New York City crossing is the first and only wo-level tube with two tracks on each level.

'ABLE 1. List of Projects Built Since 1950.

TUBE TUNNEL CONSTRUCTION

No.	Year	Name	Purpose	Location	Length	Cross Section	No. Lanes	Type
1	50	Washburn	Road	Pasadena, TX	1,500'		2	S
2	52	Elizabeth River	Road	Norfolk, VA	2,100'		2	S
3	53	Baytown	Road	Pasadena, TX	2,600'		2	S
4	56	Hampton Roads	Road	Norfolk, VA	6,900'		2	S
5	57	Baltimore	Road	Baltimore, MD	6,300'		2 x 2	S
6	59	Deas Island	Road	Vancouver, BC	2,100'		6	C
7	62	2nd Elizabeth River	Road	Norfolk, VA	3,500'		2	S
8	62	Webster St.	Road	Oakland, CA	2,400'		2	C
9	63	Chesapeake Bay Br.	Road	Norfolk, VA	5,700' 5,500'		2	S
10	67	Boucherville	Road	Montreal, QUE	2,500'		2 x 3	C
11	69	BART	Railway	San Francisco, CA	19,000'		2	S
12	73	Mobile River	Road	Mobile, AL	2,500'		2 x 2	S
13	73	63rd St.	Railway	New York City, NY	1,300'		2 x 2	S
14	74	2nd Hampton	Road	Norfolk, VA	7,300'		2	S
15	70-71	Charles River	Railway	Boston, MA	480'		2	S
16	77	Cove Point	Pipeline	Cove Point, MD	5,300'		2	S
17	79	Washington Channel	Railway	Washington, DC	1,020'		2	S
18	83	Ft. McHenry	Road	Baltimore, MD	5,300'		2 x 2 x 2	S
19	84	2nd Downtown	Road	Norfolk, VA	2,500'		2	S

Note: 1) The "Year" given is the approximate date of completion.

2) "Type" of tunnel construction: S - Steel Shell C - Concrete Shell

ublished 1987 by Elsevier Science Publishing Co., Inc.
'unneling and Underground Transport: Future Developments in Technology, Economics, and Policy
'.P. Davidson, Editor

II. PETER KIEWIT SONS' CO. TUBE PLACING EXPERIENCE

This paper will concentrate on the tubes that have been built since 1955 in which the Peter Kiewit Sons' Co. has been involved. These projects include the Deas Island Tube Project in British Columbia, completed in 1959; the San Francisco BART Tube, completed in 1969; the New York City 63rd Street Tube and Tunnel, completed in 1973; the Second Hampton Tube Crossing in Norfolk, completed in 1974; the Cove Point LNG Unloading Facility in Cove Point, Maryland, completed in 1977; and the Fort McHenry Tube Project in Baltimore, Maryland, completed in 1984. For a listing of specific data on these projects, see Table 2.

III. CONSTRUCTION CONSIDERATIONS

With two exceptions, tubes have been built through soft ground. The exceptions are the Boucherville Tube Project, where one end of the tube had to be excavated in rock and much of the rest of the trench had to be excavated in very hard dense soil, and the New York City 63rd Street Project. The rock portion of the Boucherville Project was drilled and shot in the dry from an area that was used as the enclosure or dry dock for the fabrication of the concrete tubes. The New York City tube trench excavation was done as a marine drill, shoot, and excavate operation. Tube construction can be an economical alternative to the driving of tunnels under compressed air. In the case of the 63rd Street Project, there were four bids: one was for the driven tunnel method, and three were for the tube construction method. As the labor costs of operating tunnels under compressed air go up and up, and the width requirement for the tunnels gets wider and wider, it is the author's opinion that tube construction will become a more and more popular and economical solution for crossing major water bodies. It is also feasible to construct tubes through hard dense material, including rock, although rock will certainly increase the cost of the underwater excavation.

To date, engineers have designed tubes with either a structural steel shell and a concrete lining or an all concrete shell. It is interesting to note that the first underwater highway tube, the Posey Project in California, was constructed of reinforced concrete. It is also interesting to note that many of the overseas tubes, particularly those built in Europe, have been constructed of reinforced concrete. In several instances, the concrete shells have been prestressed in order to improve their structural and watertight characteristics. The Deas Island Tube in Vancouver, British Columbia, the Webster Street Tube in California, and the Boucherville Tube in Quebec, are projects with concrete shells. The Boucherville concrete shell was prestressed.

In the case of concrete tubes, it is the practice that the units are fabricated in a specially prepared graving dock site. In the case of the Webster Street Tube in California, the dock was capable of outfitting two tubes at one time. In the case of Deas Island and Boucherville, all six or seven elements were cast within a graving dock at one time. It is possible for the owner to save depth of dredging and draft requirements by using a concrete design. However, the owner needs to consider the factor that an appropriate site be made available for the construction of the tube fabrication graving dock. The graving dock size will depend on the number of tubes to be constructed at one time. If there are three or more tube elements to be constructed at one time and the work area available permits the excavated sides to be sloped, the dock plan area at original ground level should be about two and one-half to three times the plan area of tubes under construction. There must also be storage and work area at original ground elevation, and this will run 50-100% of the dock area.

Project	Ft. McHenry	63rd St.	Trans-Bay	Deas Island	Cove Pt.*	2nd Hampton*	Chesapeake*
Year:	1980-1984	1969-1973	1966-1969	1957-1958	1974-1976	1971-1974	1959-1964
Location:	Baltimore,MD	New York	San Francisco	Vancouver, BC	Cove Pt.,MD	Norfolk,VA	Norfolk,VA
Purpose:	Road	Railway	Railway	Road	Pipeline	Road	Road
Cross Section:					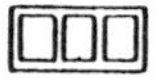		
No. Lanes:	2x2x2	2x2	2	2x2	2	2	2
Type:	Steel Shell	Steel Shell	Steel Shell	Concrete Shell	Steel Shell	Steel Shell	Steel Shell
Tube Displacement:	35,000 T	16,000 T	12,000 T	18,500 T	4,000 T	17,000 T	17,000 T
Tube Length:	5,300 ft.	1,300 ft.	19,000 ft.	2,100 ft.	5,300 ft.	7,300 ft.	5,700 ft.
No. of Tubes:	32	4	57	6	21	20	35
Maximum Depth:	102 ft.	100 ft.	132 ft.	70 ft.	60 ft.	100 ft.	105 ft.
Tide Variation:	4 ft.	5 ft.	10 ft.	11 ft.	8 ft.	8 ft.	6 ft.
Maximum Current:	2 knots	6 knots	3 knots	4 knots	2 knots	3 knots	4 knots
Foundation Placing Method:	Screeded	Screeded	Screeded	Sand Jetted	Screeded	Screeded	Screeded
$ Value (millions):	426.0	70.0	89.9	16.1	49.5	47.7	125.0

* Non-Sponsored Joint Venture

The site must be near enough to water depth that will permit the tubes to be floated out and transported to the project site with a minimum of dredging.

In today's climate where many permits, such as dredging and environmental impacts statements, are necessary to build any facility it is usually not feasible for the contractor to acquire the site and get the permits within a reasonable normal construction time. It should, therefore, be part of the owner's planning that such a site be provided. In the case of both Deas Island and Boucherville, the site was provided and the tube casting was done in a graving dock right adjacent to the centerline of the project.

The longest tube project built to date is the San Francisco BART Project, which was 19,000 feet long. This is also the deepest project, with a maximum depth of 132 feet. The New York City 63rd Street Project and the Boucherville Tunnel Project probably set the record for being built in bodies of water with the highest velocities, with the New York Project river velocities being in the range of six knots.

The length of individual tube sections usually runs in the 300-500 foot range. A tube section can displace from 4,000 tons up to 34,000 tons, as in the case of the Fort McHenry Project. The maximum width for an element is the Boucherville tube which is 120 feet. Tube height is a function of use and type of design chosen. A typical double steel shell highway crossing is usually in the 35-38 foot height range. In the case of Deas Island (see Figure 1) and Boucherville, both concrete tube, heights were only 24 and 26 feet, respectively. In the case of the 63rd Street Project (see Figure 3) which was an over-under, side-by-side configuration, total height was 40 feet. The Fort McHenry Project (see Figure 4) was a steel shell, two barrel, side-by-side tube with an overall height of 42 feet. The BART tube (see Figure 2) was of a single steel shell design and was 21.5 feet high. As modern safety standards become more and more demanding on the acceptable width of a highway, the height

Figure 1. The Deas Island tunnel.

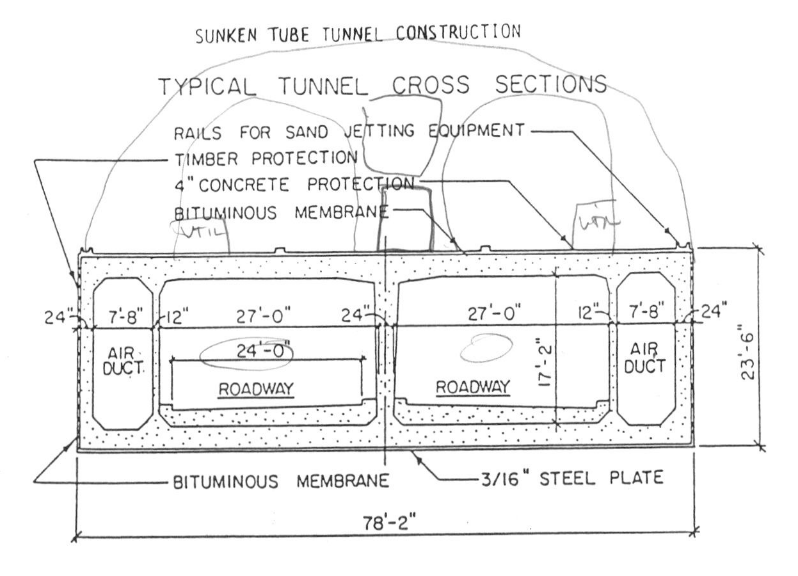

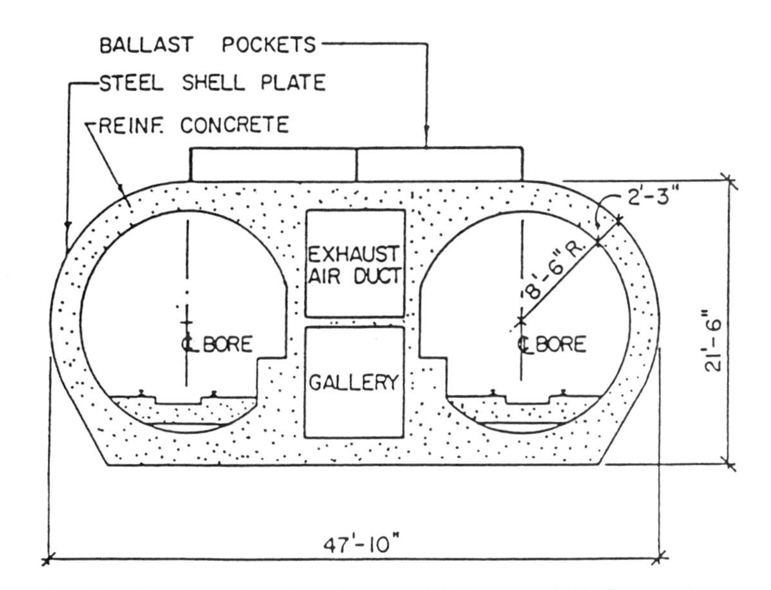

Figure 2. The San Francisco Bay Area Rapid Transit (BART) tunnel.

Figure 3. The 63rd Street Tube tunnel.

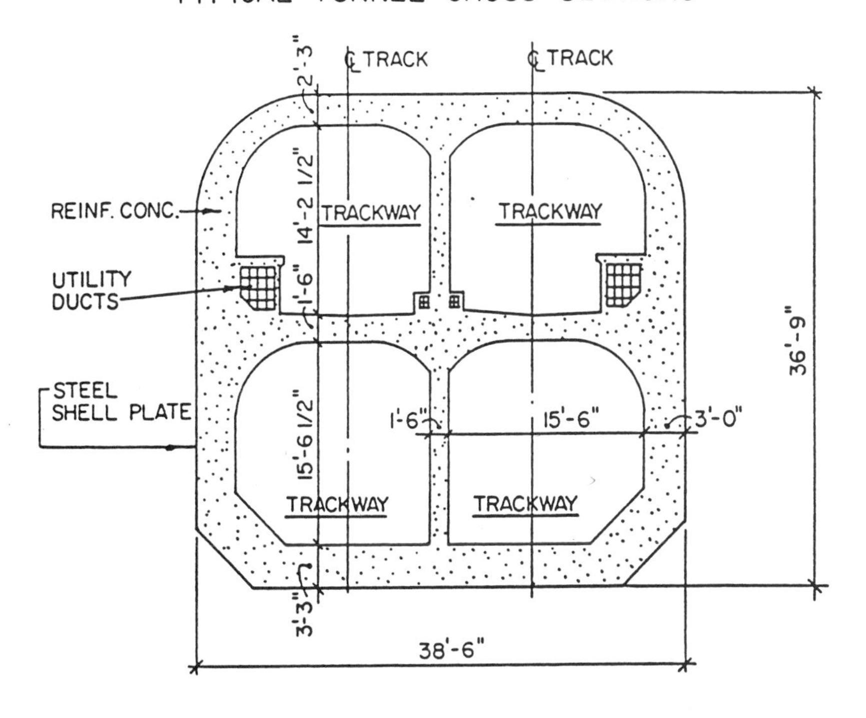

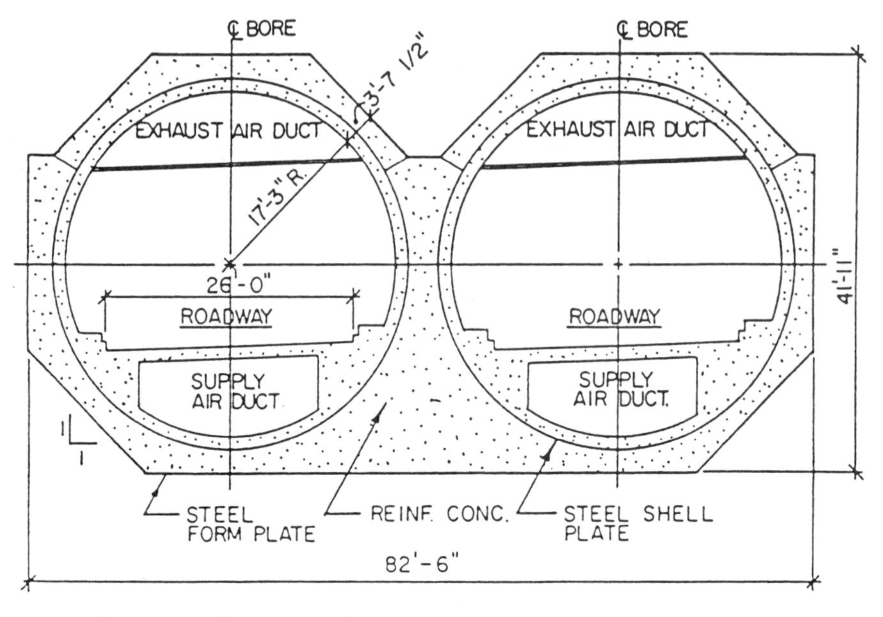

Figure 4. Fort McHenry tunnel.

of tubes built in the round steel shell configuration is going to get greater and greater. This increased height will present more and more construction problems. Most harbors are not equipped to handle drafts of more than 30-35 feet. Drafts greater than this for tube elements present either major dredging problems from the outfitting site to the tube-placing site, or require that the tubes be brought to the placing site with the concrete not yet completed, requiring the remaining concrete work to be done at the placing site. Either of these additional operations increase the risk and costs significantly.

The method used to place tubes started out to be controlled by the type or class of equipment available. This usually meant that the tubes were lowered by floating crane barges which permitted a negative buoyancy of 100 tons or less. This negative buoyancy is usually obtained by the addition of tremie concrete to the tube or with water and sand ballast. As tubes become larger and larger in displacement and the effect of changes in water density due to salinity changes or suspended solids became more and more critical, the required tube lowering weight became greater and greater. In some cases, where fresh water overrides lower salt layers, a tube can experience an apparent loss of buoyancy of up to 2.5% of its displaced weight. On Peter Kiewit Son's Co.'s first project at Deas Island, they were faced with this problem and very limited capacity tube placing equipment. They used a special placing barge constructed from four normal flat deck barges connected together with trusses. With this design, the tubes could only be lowered with about 100 tons of negative buoyancy. Since the tube element would lose more weight than this during the lowering process, Kiewit had to have a method of adding ballast as the tube was lowered. This additional ballast was added by pumping water into tanks inside of the tubes. This is a rather expensive and risky way of handling the problem and Kiewit has not used it since.

On many projects, the design engineers provide that lowering ballast weight be obtained by placing tremie concrete between the inner steel shell and the outer steel form plate. This is done after the tube has been delivered to the placing equipment. On several recent projects the tube has been designed so that it has a minimum of positive buoyancy when all concrete work has been completed at outfitting. The tube is then delivered to the placing equipment with a small positive buoyancy which can result in a free board two feet or less. The tube is then ballasted with gravel placed on top of the tube in a ballast box provided for that purpose. This is less risky, more economical, and substantially faster than tremie concrete. After the tube has been placed on the bottom, additional gravel ballast can be placed on top of or around the tube, if necessary or desired. When the tube interior work is done, additional concrete work can be performed inside the tube to give it sufficient negative buoyancy for the permanent condition. It is usual to leave such items as ventilation, curbs, walkways, and asphalt overlays to be done after the tube is in place, and with proper design this additional work can give the project its desired final negative buoyancy. Of the 19 projects completed since 1950, 6 were placed with cranes, 11 with special placing barges, and 2 with special pile-supported placing equipment. The special placing barge is frequently two long barges in the 300-350 foot length range linked together in a catamaran fashion. Such equipment can be designed and built to place the tube at any desired negative buoyancy, but the range of 400-600 tons is probably the most common. There have been several projects where the tubes were placed with substantially greater negative buoyancy than 600 tons and this has been done by designing a heavier duty lowering system. In general, the lowering system has been winch powered, but on at least two projects, the tube lowering was performed with a hydraulic jack and cable arrangement.

In the case of the steel shell design, it is necessary that the tube elements be fabricated in a shipyard. Tubes are usually launched sideways, although in the case of the San Francisco BART tube, the tubes were launched lengthways. Lanuching lengthways worked quite satisfactorily. It is necessary to make a launching engineering analysis and it may be necessary to do some reinforcing of the steel shell in order to provide adequate rigidity and strength during launching. The work of preparing the shipyard site for tube fabrication, assembling all of the special jibs and equipment necessary for the fabrication of the tube, and then the actual construction of modules, assembly of modules on the ways, and then launching the first tube can take up to twelve months. This time will vary depending on the amount of work that has to be done in the shipyard to prepare the facilities for tube fabrication and the size of the tube elements to be fabricated and launched. As the elements get longer and longer, and wider and wider, the lead time will get longer and longer.

Consideration must also be given to provide sufficient strength in the shell so that they are seaworthy enough to be towed some distance. In two cases, the tube elements were fabricated over 1,000 miles from the construction site. More common, the tube fabrication site is within 200-400 miles from the tube project. Even in many of these cases, it has been necessary to tow the elements in the ocean from one harbor to the other in order to get from either the fabrication site to the outfitting site or from the outfitting site to the placing site. Tubes have been towed both partially outfitted with only keel concrete and completely outfitted.

Because steel shelled tubes have the concrete lining constructed through a half dozen to a dozen 4-foot by 8-foot access hatches in the top of the tubes, it has been the industry's practice that as many things as possible, such as reinforcing steel, embedded items, pipe, electrical conduits, and other such items be installed at the shipyard because of much better access prior to the end bulkheads being put in place.

Foundation preparation for tubes can be divided into two basic types:

(a) a screeded gravel bed foundation.

(b) a pumped sand foundation.

There are a few variations to the above two basic methods for conditions where very unique soil must be treated or improved by the use of a pile foundation or other systems. Of the 19 tube projects constructed since 1950 in North America, 4 have used the pumped sand foundation and 15 have used the screeded bed or a pile-supported variation thereof. Pile foundation treatment has been used on three projects. The pumped sand foundation method was used at Deas Island, Boucherville, Webster Street, and Mobile. The screeded bed or pumped sand foundation method each work equally well for both concrete and steel shell tube designs. The Fort McHenry Tube is the only project for which the contractor was given the alternative of bidding on either the pumped sand foundation method or the screeded bed method. Both bids submitted for the project were based on using the screeded bed method and in the case of our analysis, there was a minimum of $3-4 million savings in the screeded bed method. It would appear from the results of the building on the Fort McHenry Project that the screeded bed method is the most viable and economical way to prepare tube foundations.

Finally, other construction design considerations are soil conditions as they affect dredging, slope stability of the excavated trench, and foundation bottom conditions as they affect the support of the tube. Variables that affect water conditions at the site are: the tide range, velocities of the current, variation of velocities and slack times, water densities and variations of those densities from top to bottom of the body water, and the variations from day to day or season to season of all of the above. Availability of outfitting sites and depths of access riverways and channels are of major importance as well.

IV. SUMMARY

In discussing the methods of outfitting, screeding, placing, and backfilling of tubes, it is apparent that a great deal of specialized and unusual equipment is necessary for a tube construction project. The development, design, and construction of this equipment takes a major, concerted effort at the beginning of any tube project. This effort will usually consume the first year that is also used by the tube fabricator in building the first tubes. A very large engineering staff is required to work up all of the designs and do the detailing of the equipment so that it can be fabricated in time to maintain the schedule. It is sometimes necessary to call on consulting engineering companies for help in order to complete the work on time. The pieces of equipment being designed and developed are normally a one-of-a-kind variety, and there are frequently bugs or problems that develop after the equipment is put into use. This requires additional engineering and time to correct.

In the case of projects with unusual and critical water conditions, it is necessary to have special studies made in order to develop proper drag and anchor loads for all of the equipment. In New York, this was very critical due to the high velocity of the East River. The joint venture was very fortunate in being able to work with the United States Corps of Engineers Laboratory at Vicksburg, Mississippi, in running these tests. They built a model of a two-mile section of the East River to a scale of 1 to 100. With this model, they were able to study and determine the drag forces of the placing equipment and of the tubes at each stage of tube placement. They were also able to determine the effect of the current on the tube once it had been placed in the trench. An added complicating factor in New York was that in order to meet the flat railroad grades and yet not get too deep with the underwater trench, the tubes had to extend above the normal river bottom. The Vicksburg Laboratory also ran many special studies in order to evaluate whether the river velocities were great enough to disturb a tube once it had been placed.

With what has been reviewed here, it may be seen that sunken tube tunnel construction can be a viable solution for either rail or highway water crossings of almost any size, length, or depth.

Buckskin Mountains Tunnel and Stillwater Tunnel: Developments in Technology

KENNETH D. SCHOEMAN
Head of Tunnel Section, Division of Dam and Waterway Design
U.S. Department of the Interior, Bureau of Reclamation
Engineering and Research Center, Lakewood, CO 80228

INTRODUCTION

Tunneling technology leaped forward with the use of the modern Tunnel Boring Machine (TBM). The first use of the TBM on a Bureau of Reclamation job was in 1964 in the Azotea Tunnel. Two TBMs were used with bore diameters of 12.5 to 13.3 feet. A very respectable average excavation rate of 55 feet per calendar day was achieved on the 67,000-foot-long tunnel. Boring of the 20-foot-diameter, 10,000-foot Navajo Tunnel No. 1 took place at about the same time and attained an average calendar day rate of 37.1 feet. Both tunnels were bored in sandstone and shale with rock compressive strengths to 8,500 lb/in^2.

Boring harder rock as in Buckskin Mountains Tunnel met the next natural challenge. In addition, the contractor overcame adversities such as caving ground while using the TBM. Completing Vat Tunnel and Hades Tunnel with ingress of large amounts of water while using a TBM also advanced the technology. However, water problems will be the subject of a future Bureau paper. Neither Buckskin Mountains Tunnel nor Stillwater Tunnel contractors experienced problems with large waterflows during construction.

In Buckskin Mountains Tunnel and Stillwater Tunnel other problems were encountered that demanded innovative solutions and these are explored in this paper. Also of interest are the new concepts and technology developed to minimize costs from the outset of these two projects. These also are enumerated here and examined, and illustrated in Figures 1-25.

BUCKSKIN MOUNTAINS TUNNEL

The Buckskin Mountains Tunnel is a 6.9-mile-long, 22-foot finished diameter, free-flow water conveyance tunnel. It begins near the Bill Williams' arm of Lake Havasu. Lake Havasu is created by Parker Dam on the Colorado River on the Arizona/California border. Water is lifted 800 feet by Havasu Pumping Plant to the inlet of the tunnel. The water thus starts its 190-mile trip to Phoenix, Arizona, via the Granite Reef Aqueduct. This aqueduct was referred to in a recent article in Engineering News Record as the "canal with cruise control" because of its automation feature.

The Buckskin Mountains Tunnel boring machine arrived at the jobsite in western Arizona in March 1976. It had a bore diameter of 23.5 feet and had capability to bore the variable hard to soft volcanic rock (andesite, agglomerate and tuff) that was known to have unconfined compressive strengths as high as 43,000 lb/in^2. That strength is ten times that of ordinary concrete.

With only about 2,000 feet of the 6.9-mile tunnel bored, a serious problem was encountered--large blocks of rock coming down out of the crown before they could be supported. The disc cutters originally protruded some 16 inches from the face of the domed shaped cutter head. In order to alleviate the problem of rock fall, the protusion was reduced to 4 inches

Published 1987 by Elsevier Science Publishing Co., Inc.
Tunneling and Underground Transport: Future Developments in Technology, Economics, and Policy
F.P. Davidson, Editor

Figure 1. Temporary post supporting blocky rock above TBM shield.

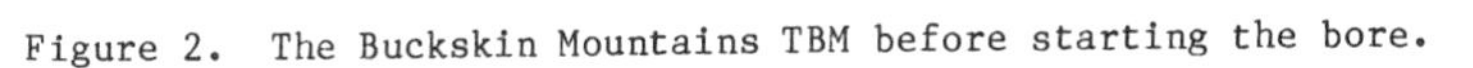

Figure 2. The Buckskin Mountains TBM before starting the bore.

Figure 3. TBM after hole through (note faceplate).

by addition of a faceplate. The exacting installation of the faceplate carried out underground proved successful and allowed the machine to proceed well through blocky ground by minimizing rock loosening in the crown.

The Buckskin Mountains Tunnel lining design and construction also incorporated another successful technological advance. The 22-foot, finished diameter lining consisted of thin precast concrete segments, four to a ring. The 5,000 lb/in^2 compressive strength concrete was variably 6 or 7 inches thick, and lightly reinforced with welded wire fabric. The welded wire fabric reinforcement provided bending strength in the segments to minimize damage from loads imposed by handling and installation. The joints were unbolted tongue-and-groove type. The top and bottom segments of trapezoidal shape in developed plan view and the side segments of parallelogram shape allowed the top segment to be placed somewhat as a large wedge.

An annular space of about 3 inches between the outside of the segment ring and the excavated rock allowed for a small amount of room to maneuver the segments into place and also allowed room for the segments to be installed under the TBM roof shield. This annular space was then filled with pea gravel as the shield was retracted, to provide rock support. The pea gravel was later grouted to provide a nearly uniform radial load for the thin segmental ring. Grout under the invert segment was kept close to the heading to provide for support of construction loads. A short way back, the remainder of the annulus already filled with pea gravel was cement grouted.

Another technical advance on the project featured the use of specially developed, durable, one component, polyurethane calking compound to seal joints and thus prevent escape of grout and insure against excessive loss of transported water by exfiltration.

Figure 4. The contractors segment manufacturing plant near the outlet portal.

Figure 5. The completed precast concrete segments in the storage yard.

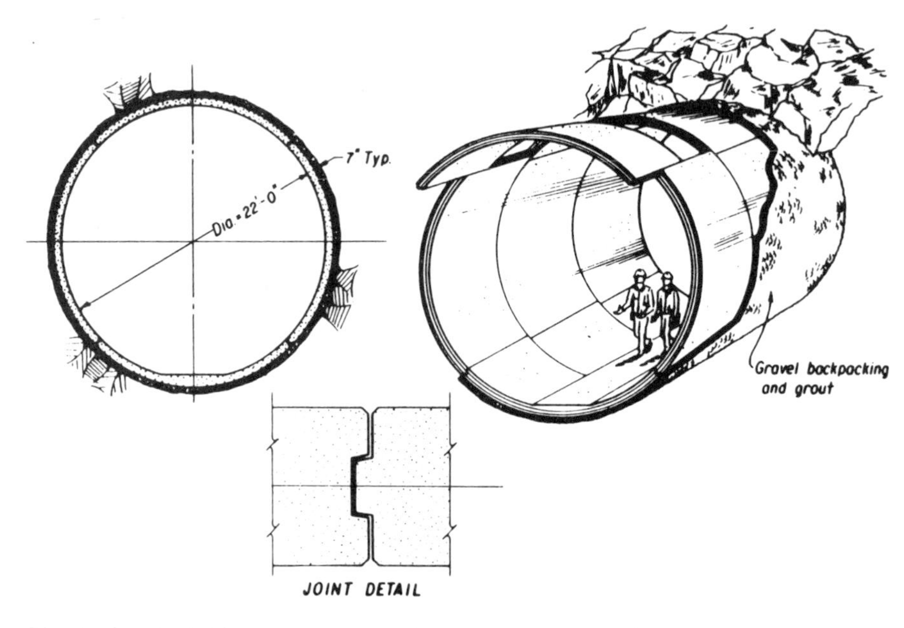

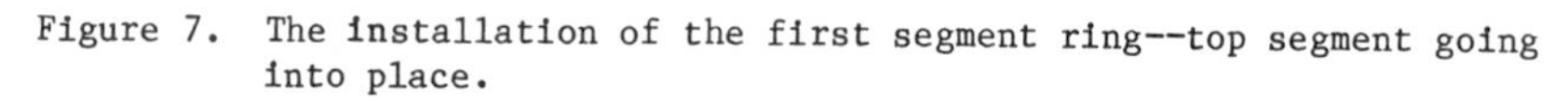
Figure 6. Artist's conception of the precast support/line systems.

Figure 7. The installation of the first segment ring--top segment going into place.

Figure 8. Conveyor belt used to load the muck train.

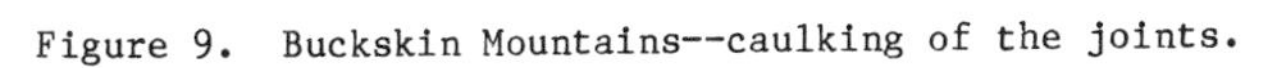

Figure 9. Buckskin Mountains--caulking of the joints.

At the end of March 1978 the TBM was stopped by caving ground. The caving ground consisted of decomposed andesite in a soft, moist state. As the machine was backed up to get to the problem area, the ground continued to cave in, creating a void 26 feet long, 48 feet across and about 30 feet above the crown. Bulkheads were constructed and 600 cubic yards of pea gravel and grout were used to fill the void. Additional caving, experienced shortly after the TBM advanced, took 40-foot rebar (drilled into the rock) chemical grout and 1,200 cubic yards of pumped concrete to finally stabilize the ground and allow the TBM to advance again. Shortly thereafter, in May 1978, a third round of concrete pumping became necessary. To overcome these severe obstacles in only 2 months can be considered a technical advance in construction techniques.

Figure 10. The cavern in front of TBM created by caving ground (note grout pipe).

In April 1979, with the excavation nearly complete, another rock support problem appeared. Unstable ground again created problems in front of the machine. The machine was backed up some 50 feet, requiring removal of ten rings of segments. A bulkhead was then constructed ahead of the machine and concrete was pumped into this fourth cavity to stabilize the ground. After the concrete gained strength the TBM advanced through this reach. Then on May 25, 1979, the hole through was achieved. Figure 11 illustrates the bore rates achieved during the excavation.

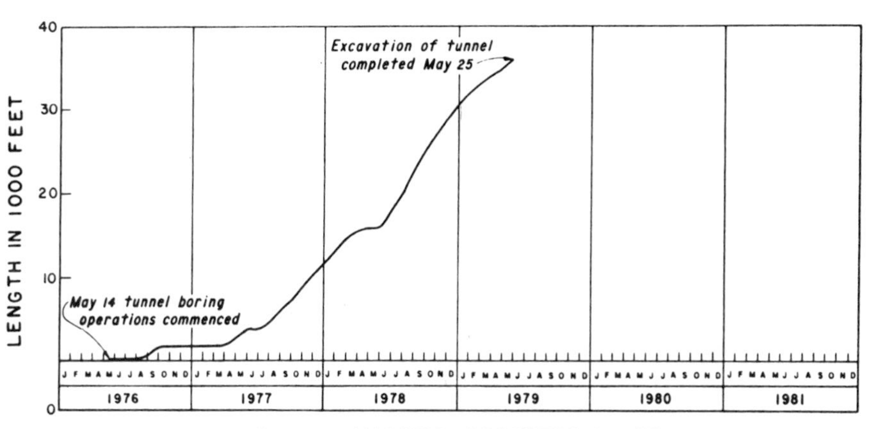

TUNNEL MACHINE—PROGRESS CHART

EXCAVATION METHOD

Robbins TBM Model, 233-172, 23 ft. 5 in. (7.16 m) diameter

EXCAVATION PROGRESS

Max. Advance/Day	120 ft. (36.58 m)
Average Rate/Day	51 ft. (15.54 m)

Figure 11. The Buckskin Mountains tunnel machine progress chart.

Figure 12. The curved portion of Buckskin Mountains tunnel with only curved utilities yet to be removed.

The now essentially completed Stillwater Tunnel in central Utah has been the subject of several articles. The tunnel is a part of the nearly complete Strawberry Aqueduct. The 40-mile aqueduct will carry water from behind the Upper Stillwater Dam, now under construction, down to the Strawberry Reservoir. There are other water diversions that feed the aqueduct and other tunnels as well that are part of the aqueduct. The completed Stillwater Tunnel is 8 miles in length, 5 miles with a 7.5-foot diameter, cast-in-place concrete lining and the remainder with an 8.25-foot diameter precast concrete lining.

Under the initial contract, as in the Buckskin Tunnel, the one pass initial support and final lining of precast reinforced concrete segments placed behind a TBM was the option chosen for construction of the tunnel, except for 1.2 miles of the inlet end, which was specified to be drill and blast with cast-in-place lining. Drill and blast excavation started at the inlet in May 1977. The inlet end reach traversed through a known major fault known as South Flank Fault without much problem.

Excavation with the original TBM with a bore diameter of 9.6 feet was started in the outlet end in early 1978. The predominant rock type for this TBM reach was expected to be Red Pine Shale with unconfined compressive strengths up to 12,850 lb/in^2 as indicated by tests on the few available borings.

Figure 13. The original Stillwater TBM at outlet portal site.

This original TBM coupled with precast segment lining having longitudinal butt joints supplied by the contractor, experienced problems as the machine became bound in the raveling and squeezing ground.

A combination of uneven gravel backpacking and high longitudinal axillary thrust against the leading edge of the in-place segments caused considerable segment damage. In September 1979, with extreme problems in advancing the TBM through an extensive shear zone, the contract was terminated by and for the convenience of the federal government.

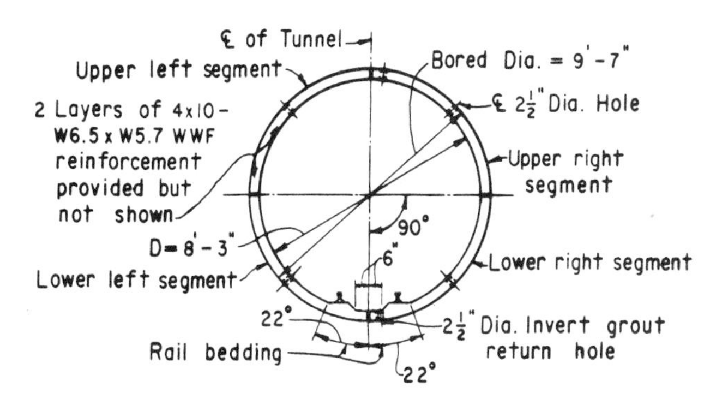

Figure 14. The precast concrete segment tunnel lining.

Figure 15. The precast concrete segment installation--first contract.

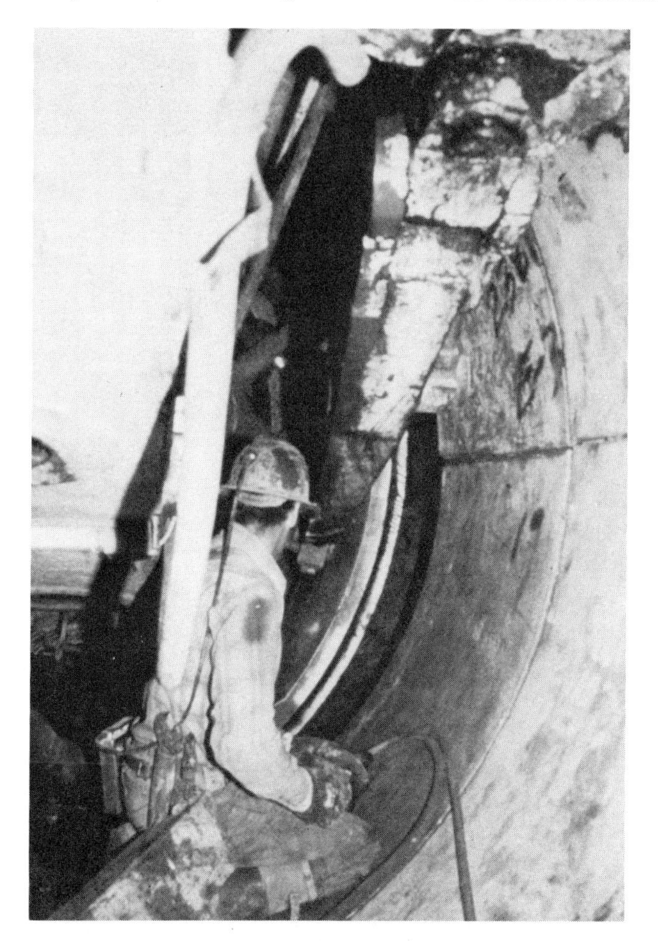

Figure 16. Segment damage.

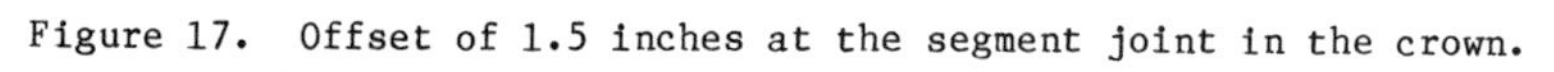
Figure 17. Offset of 1.5 inches at the segment joint in the crown.

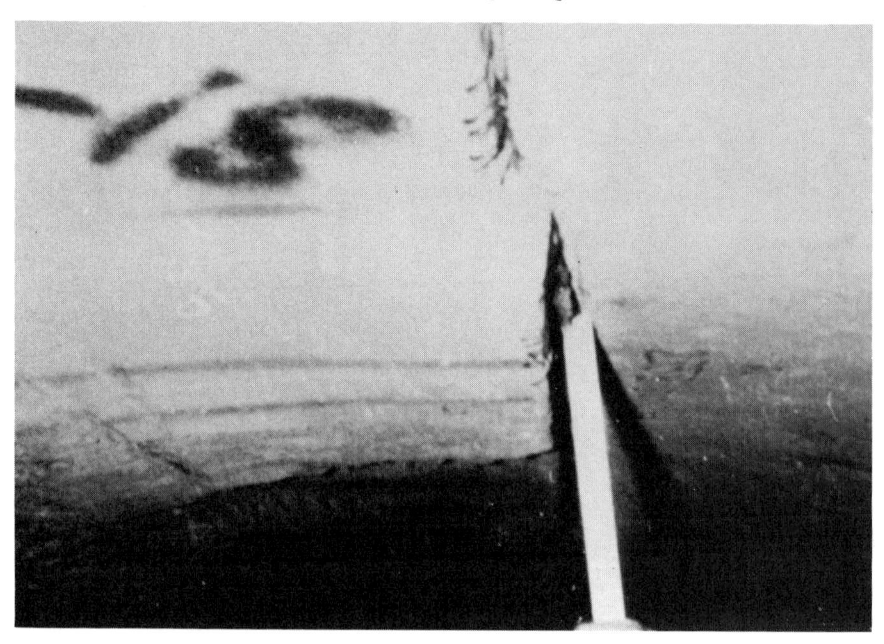

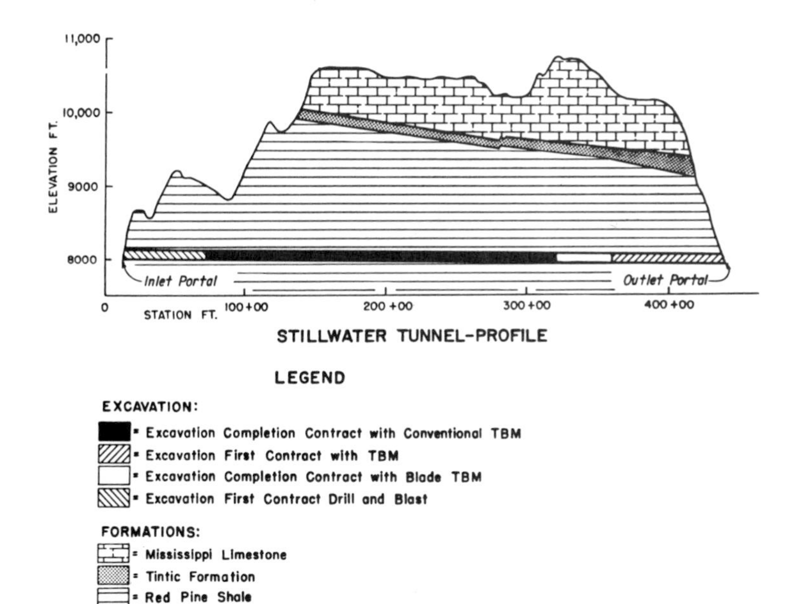

Figure 18. The tunnel profile showing geology and construction sequence.

In February 1982, a completion contract was awarded. The tunnel hole through became a reality in August of 1983. Some of the innovative methods used to bring this tunnel to successful completion qualify as developments in technology.

Timely fulfillment of the completion contract required excavation from both the inlet and outlet headings. The TBM in the outlet heading was rebuilt from a telescoping solid shield TBM into a "walking-blade-gripper" TBM. This rebuilding underground became time consuming enough; then when first completed and tested, the TBM experienced steering problems. The machine required some further modification and the addition of a Zed steering control to overcome this problem. The 12-blade grippers were usually operated in sets of six. They replaced the solid shield and were capable of applying a pressure of 90 lb/in^2 to the rock in the tunnel walls. They could travel in and out radially 12 inches, thus preventing the TBM from becoming bound in squeezing ground. Although the rebuilt TBM had a short life, boring only some 3,200 feet, the concept of the walking blade-gripper TBM was proven and may be useful for future tunneling through squeezing ground.

As mentioned in the brief description of problems encountered during the first contract, the unevenness of the gravel backpacking contributed to segment damage. In the completion contract, use of a thixotropic grout composed of sand, fly ash, portland cement, and additives took the place of gravel backpacking and a separate grouting operation, and accomplished more even filling of the annulus between rock and precast segments.

Thixotropic grout has the property of being very fluid while being pumped and will gel when no longer agitated. Thus, joints and bulkheads do not have to be completely watertight to contain this grout.

A variation of this grout was also tried. This variation makes the grout somewhat compressible by addition of styrene microballs. This

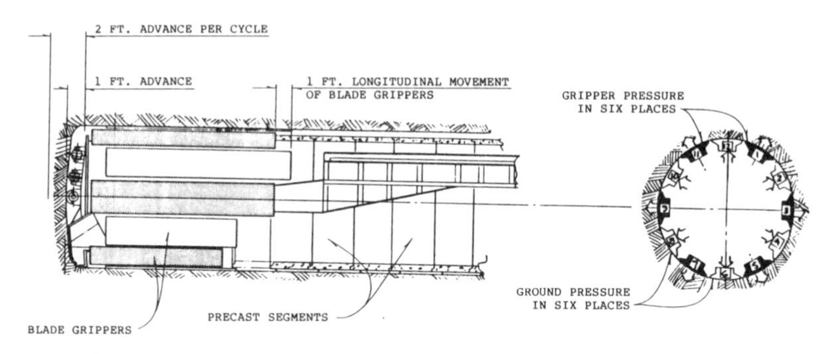

Figure 19. The "Walking Blade-Gripper" TBM concept.

Figure 20. Looking inside the "Walking Blade-Gripper" TBM in the Stillwater tunnel.

grout allows squeezing ground to move in slightly without imposing the full squeezing load unto the segment lining. After the ground is thus relieved, the load it imposes is usually permanently decreased. As a spinoff benefit, similar grout is now being used as a fill between casing pipes or tunnel linings and carrier pipes to insulate the carrier pipes from earthquake shock waves in seismically active areas.

The TBM used from the inlet end of the tunnel proved very successful even in moderately squeezing and raveling ground. The machine was a fairly open machine with a short dust shield at the cutter head followed by a finger shield under which the steel support rings and lagging were installed. The steel ribs, expanded tightly against the rock with hydraulic jacks as the shield advanced, provided good support. The modern design of this machine allowed the 10.5 foot diameter machine to be partially disassembled, brought in through the 8.5 foot wide drill and blast section and reassembled in a wide spot provided inside the tunnel. Bore rates are shown in Figure 23.

Figure 21. The TBM used to excavate approximately 5 miles from the inlet end.

Figure 22. Portion of the Stillwater tunnel complete except for the removal of utilities.

Figure 23. Record of the TBM excavation.

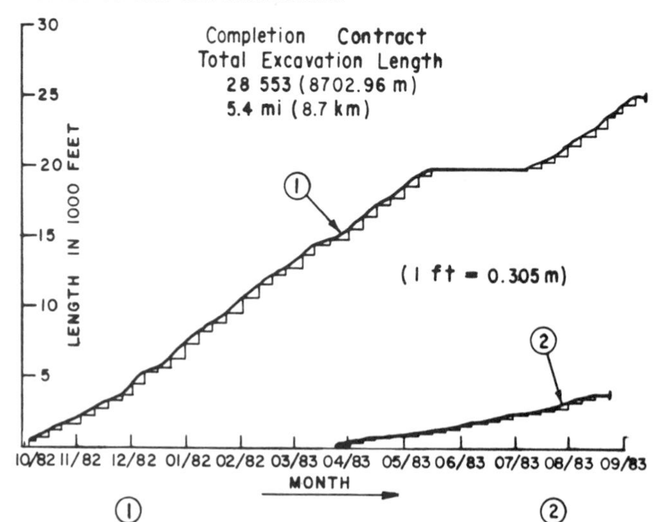

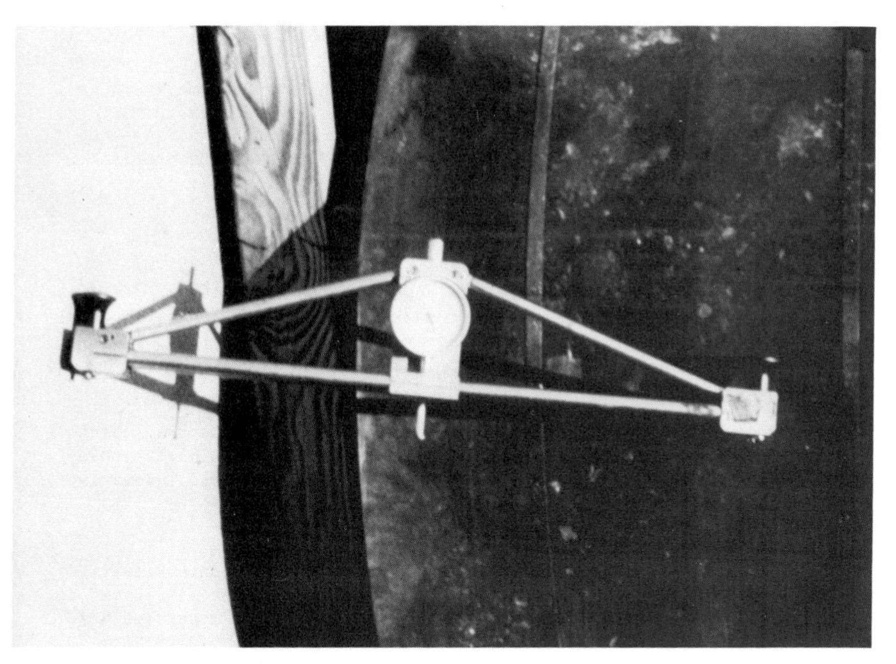

Figure 24. The curvometer/distometer instrument.

A good deal of instrumentation was used in this tunnel to measure the rock closure, rock load, and stress in both the concrete segments and steel set supports. One of the newer additions to the array of instruments used was the curvometer/distometer which measures not only the strain in the member but also the depth of curvature. Through a set of calculations, measurements obtained with this instrument are used to determine thrust and movement in the member. Thus, the actual thrust and movement experienced by the member are compared to the design thrust and rock load is also back calculated. This Stillwater Tunnel instrumentation was the subject of a paper presented by Robert A. Robinson, et al, at the

Figure 25. Incentive price share diagram.

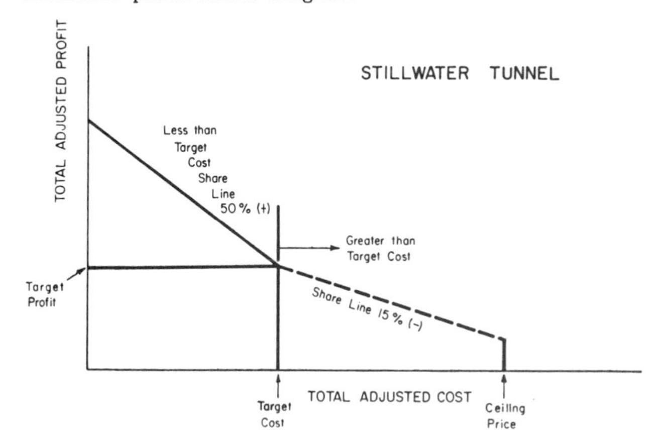

Rapid Excavation and Tunneling Conferences, held in June 1985, in New York City.

The cost plus incentive fee contract type with technical merit as well as price as criteria for choosing the contractor, was used for the completion contract. This contract type can be credited for giving the contractor added incentive to try innovative techniques in the completion contract.

CONCLUSIONS

Buckskin Mountains Tunnel: Despite adversities, substantial cost savings were realized through use of precast concrete segment lining over use of separate initial support and cast-in-place concrete final lining. The first water flowed through the tunnel and on through the nearly completed Granite Reef Aqueduct in March 1985.

Stillwater Tunnel: The big challenge was to complete the 8-mile-long tunnel through the very faulted, squeezing and raveling ground with cover to 2,600 feet. The completion contract with advanced TBMs and innovative approaches helped meet this challenge and brought this tunnel to successful completion.

These tunneling experiences substantiate that specific site related solutions are better than generalized systems. An open approach, based on all specific site data that is reasonably attainable, duly moderated with experience, is the key to success in tunneling projects. Developments in technology increase the likelihood of success and help lower project costs.

REFERENCES

Engineering News Record, 1985, CAP: The Canal with Cruise Control, March 14, 1985, pp. 34-35.

Groseclose, W. R. and Schoeman, K.D., 1976, Precast Concrete Segment Lining for Buckskin Mountains Water Conveyance Tunnel, Proceedings, ASCE, AIME, Rapid Excavation and Tunneling Conferences (Las Vegas, June 1976), Port City Press, Baltimore, Chapter 26.

Marushack, H. M. and Tilp, P.J., 1980, Stillwater Tunnel - A Progress Report, Tunneling Technology Newsletter, Number 31, September, pp. 1-11.

Robinson, R. A., et al., 1985, Ground Deformations Ahead of and Adjacent to a TBM in Sheared Shales at Stillwater Tunnel, Utah, Proceedings/Volume 1, Chapter 4.

Selander, Carl, et al., June 1980, Segmented Concrete Tunnel Lining and Sealant Systems, United States Department of Transportation, Report Number UMTA MA-06-0100-80-9.

Sinha, R. S. and Schoeman, K.D., May 1984, Stillwater Tunnel, Central Utah Project, Utah, Case History, Proceedings, International Conference on Case Histories in Geotechnical Engineering, pp. 807-815.

Underground Facilities for Defense—Experience and Lessons

LLOYD A. DUSCHA
Deputy Director of Engineering and Construction
U.S. Army Corps of Engineers
20 Massachusetts Avenue, N.W.
Washington, DC 20314

INTRODUCTION

The Corps of Engineers' historical role in underground excavation dates back to the late 1940s with conventional techniques at Garrison Dam in North Dakota and the use of a shale saw at Fort Randall Dam in South Dakota. In the early 1950s, a Mitry mole was used to excavate the power tunnels in soft shale at Oahe Dam in South Dakota. After World War II, political and economic factors changed the underground construction picture and caused a renewed interest to "think underground." As a result of this interest, the Corps of Engineers became involved in the design and construction of some very complex and interesting military projects. In addition to the military projects, the tremendous advances experienced in subsurface excavation technology in the past 20 years have impacted greatly on the traditional Corps flood control missions and the design and construction of several interesting projects such as the Hartford, Connecticut, diversion project. Although the conference program indicates the topic to be "Underground Facilities for Defense--Experience and Lessons," I must deviate a little because several of the most interesting facilities that have been designed and constructed by the Corps are classified. With this in mind, I will describe some of the more interesting details I can talk about and relate some of the major lessons learned from several military and civil projects.

Prior to the late 1960s and early 1970s and the great leap in the state of the art in subsurface excavation, conventional drill and blast techniques were almost exclusively used as the means of excavation, especially for large underground chambers.

DEFENSE RELATED PROJECTS

One of the best known large Corps projects is the NORAD underground complex in Cheyenne Mountain, Colorado Springs, Colorado. This project, originally completed in 1962 and expanded in the mid-1970s, consists of a number of rock chambers, approximately 65 feet wide by 60 feet high, excavated in granite by drill and blast methods. A considerable amount of research was performed in determining the size and shape of chambers and tunnels that would provide the required stability in the type of rock composing Cheyenne Mountain. In addition, extensive experiments were conducted in placing and loading explosive rounds to retain design configuration without excessive overbreak or shattering of the surrounding rock. Various types and lengths of rock bolts were used and patterns of placement were developed that would provide maximum protection against rock falls. Chain link fabric was installed as a protection against minor spalling and mine ties were utilized in some areas. In spite of all precautions, the usual unexpected occurred. Fault zones not revealed by test holes appeared as mining progressed, requiring reorientation of chamber areas and development of corrective measures to fit each instance. Steel sets, concrete arch sections and guniting were all used as necessity dictated. Experiments were conducted in cementation of shattered rock by means of pressure grouting with polyester and epoxy resins.

Tunneling and Underground Transport: Future Developments in Technology, Economics, and Policy
F.P. Davidson, Editor

A major fault zone made an unpredictable appearance at the intersection of two chambers; reorientation to avoid it defied all normal methods of stabilization. A massive steel column was placed in the intersection to support the rock until a positive method of support could be devised. The final solution was a dome supported on pilasters at the four corners which, in turn were supported by an inverted dome formed into the bottom of the intersection, all of massive reinforced concrete, providing, in effect, a huge, hollow, concrete ball with openings into the chambers in the form of concrete tubes until the shear zone was cleared.

Several of the most important lessons learned from the NORAD project were:

1. The immense value of pilot tunnels for large underground projects.
2. Maintaining flexibility of design during construction to best conform to geologic features.
3. Necessity of having test blasting panels to assure minimum blast damage to final rock surfaces, especially where smooth blasting or presplitting are required.
4. Use of consultants to develop blast patterns and delay intervals.

As stated earlier, there are other projects of similar scope, which I cannot identify, but which included multiple chambers up to 50 feet wide and 100 feet high using the same excavation procedures mentioned for the NORAD facility. Little has been documented on lessons learned from these projects, but discussions with personnel involved in both the design and construction surfaced one frustration that merits mention as a lesson learned. Because of the critical and unusual nature of these projects, there were often a large number of consultants and Corps staff involved during design and construction. It was very difficult to convene all of these people, on usually a very short notice, to observe critical problems and arrive at a consensus for a "fix." As a result of this experience, it is highly recommended that a line of responsibility be firmly established early in the project life for determining immediate remedial measures or changes required during construction.

CIVIL WORKS PROJECTS

In the civil works arena, the Corps has been long involved in the design and construction of tunnels, shafts, and chambers, in support of many of our water resource projects. These include such unique projects as the tunnels, underground powerplant, and lake taps for the Snettisham project in Alaska; the railroad relocation tunnel for the Second Bonneville powerhouse which was driven through the Cascade slide debris; a current project, the Spirit Lake outlet tunnel to drain the lake formed as a result of the volcanic eruption of Mt. St. Helens; and the San Antonio channel improvement which is an inverted siphon utilized to bypass flood waters.

Two projects, Hartford and Spirit Lake, produced similar results when the bid packages were offered. At Hartford, the 9100-foot Park River auxiliary tunnel was excavated at a depth of up to 200 feet below the city. This project was one of the first where the Corps facilitated the planning process by using a tunnel cost-estimating computer program. In addition, the design of the temporary and final support was assisted by use of a rock mass classification system. In the invitation three different bid options were offered: (1) conventional drill and blast with cast-in-place lining, (2) machine excavation with cast-in-place lining, and (3) machine excavation with precast lining. Out of seven bids the lowest five bidders chose the latter option. The use of precast liners for this project was the first use of this technology by the Corps. This job was completed using a fully shielded hard rock tunnel boring machine (TBM) which cuts a 22-foot, 3-inch bore.

The Mt. St. Helens debris avalanche, which blocked the natural Spirit Lake outlet, could not safely pond water above an elevation of about 60 feet below its crest. Because Mt. St. Helens and environs were declared a National Volcanic Monument, site disturbance became a key criterion, as well as stability, constructibility, and cost. These considerations led to the selection of an 8500-foot tunnel. The invitation allowed either drill and blast or TBM excavation. A range in sizes of tunnel was specified to permit the use of existing and available TBMs. There were 14 bidders, 5 submitted bids for drill and blast, and 9 for TBM excavation. The successful bidder chose an 11-foot-diameter TBM.

Lessons learned from these projects were:

1. Offering different bid options appeared to result in obtaining lower bids.

2. Allowing a range in tunnel size allowed the use of available equipment resulting in lower bids.

Other civil works Corps projects using innovative subsurface technology are the Skiatook dam in Oklahoma and the Yatesville Dam in Kentucky where a "roadheader" mining machine was used to excavate 15-feet diameter outlet tunnels in moderately hard rock. At the John Day powerhouse project, a fish transport diversion tunnel is being excavated through 4000-6000 lb/in^2 concrete using a roadheader. These machines have proven to be a highly innovative tool which can be utilized where odd-shaped excavations are required and blasting is not an acceptable option.

At the Snettisham Project near Juneau, Alaska, the Corps constructed an unlined power tunnel and underground powerhouse in 1970-1973 and is presently constructing a second unlined power tunnel. A lake tap was used to draw down a natural lake in the original work so that the intake structure to the 8600-foot power tunnel could be built "in the dry." In the present construction, a lake tap will provide discharge directly into a 6000-foot tunnel from a second lake.

The underground powerplant chamber for Snettisham was 147 feet long and 45 feet wide. Support was entirely with rock bolts to form a compression arch overhead and to prevent rock movement in the side walls.

Lessons learned from this work so far include the following:

1. The amount of explosives used in the smooth wall blasting specified for the powerhouse could be reduced below the amounts determined from the formulas then in use. This was determined during a series of experimental blasts by the contractor beginning with loadings in the perimeter holes or presplit-postsplit holes which were as specified. These loadings used trim powder cartridge with spacers for decoupling and deck loading. Each loaded hole was detonated with detonating cord. These loads were reduced until it was determined that relieved perimeter or presplit-postsplit holes could be successfully blasted using only 3/8-inch detonating cord.

2. The criteria in use in 1970 to determine tunnel lining requirements were overly conservative. The tunnel lining design assumed no restraint attributed to the rock. All but 10% of the tunnel was unlined, but the lined sections, when designed to the existing criteria, were massive and heavily reinforced. The lining design criteria were based on earlier Corps projects where shallow, less competent materials were encountered in fully lined tunnels. New analytical techniques allow an analysis which can include the restraint offered by massive igneous or metamorphic rock formations.

3. Rock bolts, if properly installed and tensioned, can provide adequate support of a large underground opening.

STUDIES FOR OTHER AGENCIES

As a federal agency and a leader in engineering and construction, the Corps of Engineers is frequently called upon by other governmental bodies to assist in many ways related to subsurface technology and construction. In support of the Department of Energy, the Corps is performing construction management through our Albuquerque District for their Waste Isolation Plant Project (WIPP) site in southeastern New Mexico, near Carlsbad. This project has been designed for the disposal of military nuclear waste. Chambers are being excavated at a depth of 2,150 feet in bedded salt utilizing a roadheader machine. The only problems reported related to this work have been difficulty in finding experienced miners and inspectors to perform quality control duties.

In connection with studies being performed for MX missile siting (deep basing), the Air Force Ballistic Missile Office requested the Corps to conduct several studies. One, a study documenting the state of the art of construction technology for tunnels and shafts and the other, a study of contracting methods for deep basing. The first study was done through the Waterways Experiment Station where the following tasks were performed:

1. A review of background data on deep-basing concepts and requirements;

2. An extensive literature search and review of tunnel and shaft construction methods;

3. Consultations with leading tunneling experts and tunneling and shaft equipment manufacturers;

4. Preparation of a state-of-the-art report.

While this report generally concentrated on conventional and/or proven methodology, it also examined briefly novel techniques such as projectile impact, laser beams, water jetting, and chemical methods. Also considered and evaluated in detail were the various TBMs and partial face tunnel machines (roadheaders).

The second study, contracting methods for deep basing, was performed by our Omaha District. Following the trend of a study done for the National Academy of Science, "Better Contracting for Underground Construction," the study reviewed current practices and legal and regulatory requirements applicable to the Department of Defense. It recommended changes in procurement regulations and methods, including better means of sharing risks, site-specific differing site conditions clauses, use of a disputes advisory board, contractor financial incentives, cost escalation provisions, escrow bid documentation, and distribution of the government's geotechnical studies and reports to the contractor. It also advised that for large projects, required equipment characteristics and performance should be specified.

RESEARCH

For many years, the Corps has been actively engaged in research in subsurface excavations, i.e., tunnels, shafts, and chambers through its involvement with various government committees, professional societies, academic institutions, and our own research facilities. Much of this research is in the area which I would like to call high risk; that is, research that has a high potential payoff but a low probability of success. I mention this to reinforce our continued interest in this area of technology and our mutual interest to solve the many problems that will require solutions when we work in this complex medium, "the underground."

CONCLUSIONS

The Corps of Engineers has and will continue to design and construct some very interesting underground projects. With this in mind and having reviewed a good deal of past Corps experience, I would offer the following observations and conclusions.

Our future research efforts should be concentrated on developing the means to know what is ahead. Can we locate and predict the presence of large amounts of water? Can we predict with some accuracy, the size effects of openings from pilot bores or borings to the final size and shape of the opening? In short, you can never have enough good information. There appears to be little promise for a payoff from the government's standpoint, to continue to look at novel drilling methods or improved equipment development. To achieve significant future savings, it must come from better contracting practices and contracting management. We should strive to implement changes in federal contracting policy (DAR and FAR) as recommended by the National Academy of Science and the Corps of Engineers study group. These changes would most certainly result in an expanded group of contractors who have in the past been reluctant to gamble on underground construction. This increased involvement would lead to more competitive pricing and hopefully lower costs. Until such changes are implemented, it will be extremely difficult to assure more equitable, efficient, and comprehensive contract documents for underground tunneling work. I would also add that we need to continue to offer experience and expertise in this type of work to our young engineers, geologists, and construction inspectors.

In the meantime, we must retain our professionalism and integrity as we deal with the so-called "underground." Only in this manner can the industry progress and provide for the resurgence of tunneling technology.

PART II

NEW TUNNELING AND TRANSPORT TECHNOLOGIES

High Tech Makes Itself Felt in Tradition-Bound Areas

Designing Geotechnical Equipment

JOHN W. LEONARD
Vice President
Morrison-Knudsen Co., Inc.
Boise, ID

This paper discusses tunneling, particularly in rock. It defends the use of tunnel boring machines (TBMs), but concludes with the idea that the original concept makes further improvements of a significant nature doubtful. It does intimate a new concept whose time has come.

Excellent TBMs for rock are available. There has been criticism by some concerning the machine's operating time. There also has been talk of the need to improve muck handling systems.

The combination of TBMs and shotcrete lining has yet to be worked out satisfactorily. It is accepted that shotcrete does stick to milled surfaces as well as blasted surfaces, but unless you employ a big diameter machine, shotcrete can delay progress. It makes an unholy mess (onto hydraulic cables, controls, drills, any orifice, dust, rebound, caustic accelerators). In addition, my co-workers and I feel that for antislaking use, nonbolted shotcrete with a TBM is of highly questionable benefit as support: so little benefit results that in almost all cases the shotcrete was probably not needed.

It takes experience, knowledge, and skill to operate a TBM. When we have these prerequisites we find the time it does operate divided by time it can possibly operate is acceptable. It is 10% or so less than the operating ratio we experience with D8 or D9 tractors ripping. CAT has made a good machine for years; it has the advantage that we can work on it in one of our own shops. I am not saying the TBM cannot be improved; it can be, but we are hitting up against economic limits.

Let us look at things that limit its possible operating time.

<u>Cutters</u>

Cutters will get 300,000 revolutions in high silica granitic material. CAT tracks in high silica sand on Tenn-Tom excavation of 1,000,000,000 cubic yards clocked just under 300,000 revolutions--and CAT does not make bad tracks. Silica has similar effects on raise drills, bits on conventional air tracks. It is a question of steel, the state of the art.

<u>Support</u>

This is a function of ground conditions: TBM use disturbs the ground much less than drill and blast. Thirty percent is a good figure for support difference; this figure would be considerably lower if young field geologists would let us bolt the ground instead of the drawing! It would be wiser to bolt the ground in accordance with actual variations in its condition, not according to a pattern shown on drawing with nice regular spacing always at normal angle to the excavated surface. Rock <u>is</u> joints; the joints' patterns and openings over the length of tunnel do <u>not</u> always look like the drawing. The lesson in all of this is to <u>bolt the ground</u>. The bolting shown on drawings is really for the purpose of estimating quantities.

Tunneling and Underground Transport: Future Developments in Technology, Economics, and Policy
F.P. Davidson, Editor

If lining is required, thickness can be reduced and often can be eliminated. Maintenance pumping is much cheaper than lining, particularly if interest costs are considered. Granted, there is a mental barrier here because "them guys" pay for lining; "we" pay for maintenance.

Other

Other considerations enter into design decisions, such as trains, whether we are building a portal or a long shaft, and the presence or absence of water. For twenty years or longer we have been trying to make conveyors or pipelines beat the use of railways in tunneling; we have not yet succeeded.

FACTORS THAT SET PRODUCTION RATES

1. Speed of Revolution of the Cutting Blades

We have increased speed by adding power to the machines. Academia now agrees with our position that increasing speed significantly beyond present levels results in offsetting decrease in the rate of penetration. If academia agrees, we must be correct! However, for those skeptics who remain research-oriented, I suggest that we explore the question by determining optimum speed for five selected rock types.

2. Rate of Penetration

There is still room for improvement here, but there are limits. Remarkably, penetration doubled every ten years from 1958 to 1978.

TBMs substitute equipment for labor. That is fine, but we are talking of up-front money--and interest rates are not low or unimportant. Delayed deliveries mean delayed excavation, and the net result is exposure of the entire operation to cost escalation. This remark is especially applicable to portal projects and new machines.

The main trouble with the TBM is in its original concept: it is the hard way. Take a 35-foot diameter TBM with 80 cutter blades gaining one-quarter inch every revolution or 48 revolutions per foot advance.

The average distance of the cutter from the center of rotation is:

$$0.7 \times \frac{35}{2} = 12.25 \text{ feet}$$

Travel distance per cutter:

$$2\pi(12.25) = 77 \text{ feet per cutter}$$

$$80 \times 77 = 6{,}160 \text{ feet of travel per revolution}$$

At 1/4 inch per revolution
= 48 revolutions per foot advance

$$48 \times 6{,}160 = 295{,}680 \text{ feet to gain one foot}$$

So in this case, contractors and manufacturers have done fairly well with a workable, but poor concept.

A new concept is needed. Where does a new concept come from? Obviously, from wisdom and experience. For years, engineers and geotechnical types have impressed on us our intellectual limitations:

"Any damn fool knows that wasn't the way to do it." .... "Why did you let that fall in?" .... "We knew you were too cheap." .... "A man's intelligence is directly proportional to his distance from the heading." So we do know where the wisdom is.

I sincerely believe that the engineers and geotechnicians, because of experience, developing knowledge, and possible improvements in exploration methods, are now capable of taking a new step to improve and apply a basic concept.

They can classify the underground feature as follows:

3. Ground

I suggest 15+ classifications of ground. Surely more than dirt or rock, and hopefully much less than 40 categories. Here is a "first cut" at it:

Water

Wet or dry (If wet, can it be unwatered, or partially lowered?)

Purpose

Transit
Liquid carrying pressure
Liquid carrying low pressure
Short time, length of use (diversion tunnel?)

Surface-settlement

Critical
Noncritical: river
open country

Length

0-1000 feet	Minor equipment
1000-5000 feet	Medium
5000 feet and over	Major

Labor rates

High - major urban areas, compressed air, working time low
Regular

Having determined the presence and weight to be accorded the above factors, the experienced engineer-manager will be ready to set forth the type of equipment to be used and the preferred method of excavation: blind shield, air or slurry, earth balance, roadheader, shield open, shield faced, or drill and blast. Note that by using this system of analysis, the decision on equipment can be made in a rational manner; it covers the bases. The tunnel can then be designed in the light of the equipment and methods specifically selected for the job at hand.

I believe the resulting savings would be significant. Unknowns are reduced. And in the light of better-organized information, a costly "conservatism" can be kept within reasonable limits. There is a much-diminished risk of bad surprises, of finding conditions radically different than were anticipated, if the method of excavation and the equipment are designated after more study and with less emotion, preferably not just a few weeks before the bid! Risk-sharing and escrow agreements have become accepted practice; so a method of reasoning which is impeccably logical should, in time, be seen as an additional precautionary step.

By whatever means, we should recognize the overriding necessity of better integration of geotechnical design and construction.

To put all this in context, I believe I can truthfully say that 90% of the rock tunnels we have been involved with did not need lining from a structural point of view. The remaining 10% of these tunnels would require lining only over very limited stretches. The tunnels stood up for months with only our initial supports before we put in the concrete lining. Why then line?

The designers may say. "We did not specify initial supports. We did not supervise installation of the bolts; the bolts must be grouted." I reply, Either "spec" it, inspect it, or look at it after all the tunnel-driving is done. If you wish to, then require the grouting of bolts and maybe even shotcrete portions of the job. The inherent trouble with concrete lining is that it requires a minimum thickness for forming and workability. For instance, for concrete pipe, there is a thickness of one inch per foot of diameter. In the tunnels I know, the lining is often 10-12 inches thick. A reduction in invert which is not formed of 6 inches in the thickness of such lining would save $30 per foot on a 20-foot-diameter tunnel.

If we are to put the economics of tunnel building "front and center," we <u>must</u> study the ground first--tunnels are not built in the air, at least not usually--and then we must select excavation methods and equipment in an orderly manner, on a case-by-case basis and not because of preconceptions or built-in preferences for particular types of machinery. Over-routinized procedures lead me to recall the older prewar military saying, "Society can't afford us any longer." <u>Per ardua ad terra</u>.

Dr. David Saxon, Chairman of the Corporation of the Massachusetts Institute of Technology, welcoming guests to the conference dinner, held on April 24, 1985, in the ballroom of the Copley Plaza Hotel in Boston, Massachusetts. Dr. Saxon, in his remarks introducing Dr. Harrison Schmitt, emphasized the criticality of a properly prepared space program and of adequate technical training for participants in each of the many technological domains necessarily involved.

Thomas Stockebrand, center, receiving the congratulations of Dr. Harrison Schmitt (right), the geologist-astronaut who piloted the lunar module and participated in the Apollo Program's first manned expedition to the moon, and of Ted Johnson, International Marketing Director (retired) of Digital Equipment Corporation. Mr. Stockebrand was awarded the 1985 Isambard Kingdom Brunel prize for his successful demonstration of supersonic tube flight at the conference. Mr. Johnson had recruited Mr. Stockebrand for the task of designing, testing, installing, and operating the model.

Sir Robert G.A. Jackson, senior advisor to the Secretary-General of the United Nations. As the naval defender of Malta in World War II, Sir Robert optimized reliance on tunnels and underground space.

At the conference dinner in the Copley Plaza Hotel, John W. Landis (second from left) chatting with Dr. Frank P. Davidson (left) and with Dr. David Saxon and Dr. Harrison Schmitt. Mr. Landis served as co-chairman of the conference. He is President of The American Society for Macro-Engineering and a senior vice president of Stone & Webster Engineering Corporation.

Dr. Peter Glaser, Vice President of Arthur D. Little, Inc. and a co-chairman of the conference, with Tadashi Ashimi, Director of the Local Planning Division, Comprehensive Planning Bureau of the City Government of Osaka, Japan.

An informal moment at the conference. Professor Manabu Nakagawa standing (left) and Professor Lester C. Thurow of the Sloan School of Management, chat with Dr. Herbert H. Einstein, Professor of Civil Engineering and a member of the United States National Committee on Tunneling Technology (National Research Council, National Academy of Engineering, National Academy of Sciences). Professor Nakagawa is Chairman of The Japan Institute of Macro-Engineering; he holds a Chair of Economic History at Hitotsubashi University, Tokyo.

Hans Boesch, Construction Manager of the J.A. Jones Construction Company of Charlotte, North Carolina, in conversation with Kaj Havnoe, Executive Vice President, Technical Department, of Christiani & Nielsen A/S, of Copenhagen.

Thomas R. Kuesel, Chairman of Parsons, Brinckerhoff, Quade and Douglas (New York consulting engineers).

Dr. Robert Salter, originator of the "Planetran" concept of a world-girdling supersonic subway, in conversation with Dr. Ted Okumura of Shimizu Construction Co., Ltd., of Tokyo and Los Angeles.

William Stevens, President of Taurio Corporation, with Janet Caristo, President of Macro-Projects International, Inc., and William Bohannon, a specialist on the Taurio staff with an interest in midchannel access technology.

Professor A. George Schillinger, Dean of Management at the Polytechnic University of New York (left), with Dr. Nancy Needham and Dr. Harold Luria, Dean of Engineering at Northeastern University, Boston.

Professor David Gordon Wilson, inventor of the Palletized Automated Transit (PAT) system, at table with Ms. Sara Jane Neustadtl, author of Moving Mountains (AMC Press) and Thomas Stockebrand of the senior scientific staff of Digital Equipment Corporation in Albuquerque, New Mexico.

Bankers in a corner of the banquet room: seated are Ian Wilson and Malcolm Binks of Merrill Lynch Capital Markets. Standing, left to right, are Alfred E. Davidson, former Director of the Paris office of the International Finance Corporation (a division of the International Bank for Reconstruction and Development, Washington, D.C.), and John Bennett of the National Westminster Bank of London. "Natwest" has been a leader of the finance consortium for the railway tunnel under the English Channel.

Professor Jean-Claude Huot of Concordia University, Montreal, chatting with Dr. Nancy Needham, Vice President of The American Society for Macro-Engineering. Professor Huot chaired an international colloquium on mega-projects held in Lyon, France, September 12-15, 1983, sponsored by the Mayors and Chambers of Commerce of Montreal, Canada, and Lyon, France. Dr. Needham, a former member of the faculty of the Harvard Graduate School of Business Administration, now heads the Institute for Corporate and Government Strategy and serves, in addition, as Adjunct Professor at the Polytechnic University of New York.

Dr. Peter Glaser (right), co-chairman of the Conference, introducing the "mock hearing" panelists at the concluding session: Col. Carl B. Sciple, Commander and Division Engineer, New England Division of the U.S. Army Corps of Engineers; Charles B. Steward, Assistant Director of Construction, Massachusetts Bay Transportation Authority (MBTA); Walter J. Hickel, former Governor of Alaska and onetime United States Secretary of the Interior, now Chairman of Yukon-Pacific Corporation, and Paul E. Tsongas, former United States Senator from Massachusetts and now a partner in Foley Hoag & Eliot, Boston.

William Kuhlmann, President of Taurio Corporation of Groton, Connecticut, together with William Bohannon and J. Vincent Harrington, the two Taurio engineers who presented a paper and physical model of the midchannel access system for building intercoastal tunnels. To their right is Joe Benjamin, technical delegate of Wirth GMBH of Erkelenz, Federal Republic of Germany.

Paul Tsongas, former United States Senator and a firm advocate of placing hazardous industrial operations on artificial offshore islands.

Governor Walter J. Hickel addressing the final session of the conference. Governor Hickel's interest in conduits, whether subterranean or on the surface, led him to advocate a gas pipeline from the north of Alaska to Valdez, a port in the southern portion of the state. Governor Hickel's Yukon Pacific Corporation is seeking authorization to ship petroleum products to Alaska's "natural markets" in the Far East; the United States government is urged to contract with Mexico for the counterbalancing import of oil.

Professor David Gordon Wilson exhibiting a working model, crafted by an M.I.T. student, Brian Barth, of Palletized Automated Transit (PAT). The system, patented by the Massachusetts Institute of Technology and developed with funds provided by the United States Department of Transportation, would have several stipulated advantages: it would be "fail-safe" and would therefore reduce the toll of highway deaths and injuries; it would provide seven to ten times the capacity of a normal road lane and would therefore reduce the dimensions and hence the costs of building tunnels, bridges, and roads utilizing the system; and because power would be provided from a central generating plant, air pollution could be correspondingly mitigated.

Professor Lester Thurow of M.I.T.'s Sloan School of Management, pointing out the essentiality of speed of construction if financing costs are to be manageable.

Constraints on Tunneling Technology

EDWARD J. CORDING
Professor of Civil Engineering
University of Illinois at Urbana-Champaign
Urbana, IL 61801

One of the things I am going to be discussing in this chapter is how difficult it is to evaluate ground conditions from the geotechnical standpoint, because that is my background. I do not want to be too negative about it because we have improved quite a bit in our ability to evaluate ground conditions and to explore. However, there are still quite a few difficulties in obtaining precise information as to how the ground will perform. That is the cause of some of the risks in tunneling and some of the problems that do develop, because we cannot anticipate everything that is going to take place on a project.

Figure 1 is an outline of the limitations of current tunneling. Depending upon who is giving the lecture you will find that there is quite a difference in what a person considers limitations. However, I decided, in the last few weeks that the title should be changed. I would like to talk about the constraints that are involved in tunneling, and so I did change the title to "Constraints." We can look at the constraints, as a civil engineer might, in terms of three different categories: first, the strength of the material, second, the loads that are applied to the tunnel and the reaction of the tunnel itself to these loadings, and finally the boundary constraints or geometry. The constraints may be beneficial, permitting us to do things that cannot be done with the above-ground structure, or they may make it more difficult for us to work underground.

1. MATERIAL

- CAN NOT SPECIFY
- DIFFICULT TO SAMPLE OR OBSERVE
- DIFFICULT TO PREDICT

2. LOAD

- ALREADY IN PLACE
- HIGH
- UNLIMITED BUTTRESS
- REDUNDANT STRUCTURE

3. GEOMETRY

- LONG-LINEAR-LIMITED ACCESS
- PROGRESS
  - CONTROLLED AT TUNNEL FACE
  - DEPENDENT ON PERFORMANCE . . . .
    - OF A FEW INDIVIDUALS
    - OF A MACHINE THAT OCCUPIES ENTIRE FACE

Figure 1. The limitations of current tunneling technology.

Tunneling and Underground Transport: Future Developments in Technology, Economics, and Policy
F.P. Davidson, Editor

So these constraints are real physical conditions with which we must work. The unique combination of the constraints in three categories are what I think attracts many of us to the field of tunneling. Reviewing them may provide some insight into the current status of tunneling technology as well as the progress and innovation that might be achieved in the future. We need to recognize the conditions with which we have to work in order to do our best work. There are other constraints that are organizational, which will require some adjustments over the years ahead, in order for us to obtain maximum benefit from emerging tunneling technologies.

First, I would like to look at the material. The material is natural. We are working with rock, soil, and water. It has been said that if it were not for water, the discipline of soil mechanics would not be needed. Perhaps a corollary might be that if it were not for soil, we would not have difficulty in rock. Most problems in rock are when it is not.

Another situation that is very difficult is mixed ground in which combinations of rock and soil are present, making it difficult to advance the tunnel. That is where many of our difficulties now lie, in the ability to support the ground and to advance the machine effectively, to continue the rate of advance and not get stalled. Figure 2a is an example of a changing ground condition. This is not strictly a mixed face condition, in which you encounter two different materials in the face at the same time, but is a situation in which a tunnel shield that is set up to operate in sand encounters a different soil condition--a hard clay, as it advances. The example illustrated is an open digger type of shield with the sand maintaining a natural slope, or angle of repose, in the tunnel face. A shelf is placed at the front of the shield allowing two sand slopes to form thus reducing the amount of sand that must be maintained in the face to prevent soil from running into the tunnel. In other words, by putting in multiple shelves you can maintain a steeper front and still have the sand at its angle of repose, acting to stabilize the face. As the shield enters the hard clay, the clay is prevented from squeezing into the tunnel by the shelf, and the shield cannot be advanced efficiently, so the shelf is removed. Later, as the shield breaks from hard clay into sand again, the shelf is not in place, and suddenly there is a condition in which a run or a flow might occur and even result in development of a sinkhole to the ground surface. So changing, or mixed soil conditions are often difficult. Tunneling in a situation in which buried valleys filled with sand pop in and out of the top of the profile is difficult, particularly if the sand is ponding water, which would be difficult to locate and effectively dewater with a general dewatering program.

Figure 2b is another example of a mixed-face condition, in which the tunnel shield being advanced in sand encounters large boulders or rocks. When the rock or boulders are encountered, they often cannot be brought into the machine, be broken and excavated without steepening the tunnel face and, in the process, causing the sand to run or flow (if water is present) into the tunnel. Although a "digger" type shield is illustrated in Figure 2, similar--even more severe--difficulties may be encountered when using a shield that has a rotating head or "wheel" to cut and excavate the mixed ground. Large loss of ground and ground settlements can result. Often, the material in which we work is variable and difficult to control and advance through.

As outlined in Figure 3, the material--rock, soil, water, and mixed ground--must be accepted, we cannot specify what the material will be, although we may try to avoid it or even modify it. If we have to accept

MIXED OR CHANGING GROUND

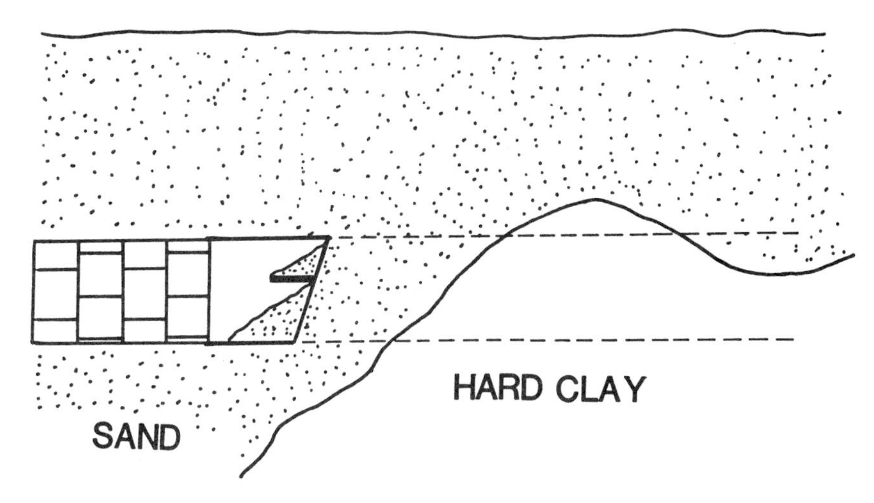

Figure 2a. The mixed or changing ground: Sand, hard clay.

Figure 2b. The mixed or changing ground: Sand, boulders, and rock.

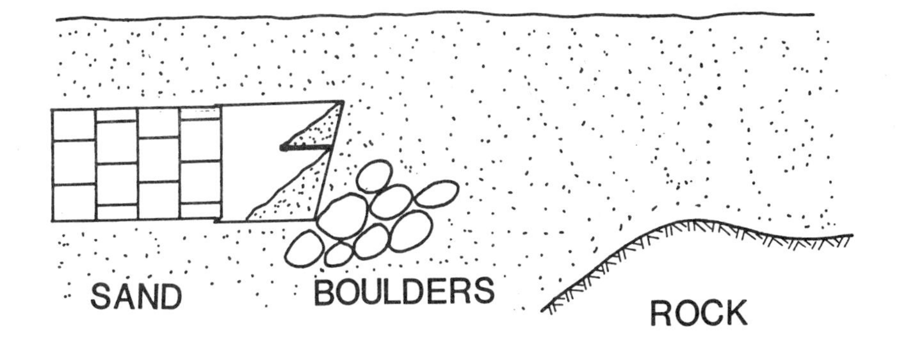

1. **MATERIAL** ROCK-SOIL-WATER
MIXED GROUND

- CAN NOT SPECIFY
  - AVOID OR COPE
  - EXPLORE-OBSERVE
- DIFFICULT TO SAMPLE OR OBSERVE
  - OPAQUE
  - ACCESS DIFFICULT
  - SAMPLE SIZE-NOT REPRESENTATIVE
  - SAMPLE DISTURBANCE
- DIFFICULT TO PREDICT
  - GROUND CONDITIONS
  - GROUND PERFORMANCE
    - TEST LIMITATIONS
      - INDEX TESTS
    - MECHANISMS OF FAILURE
    - INTERACTION OF MATERIAL, LOAD, GEOMETRY, AND CONSTRUCTION PROCEDURE
- REVEALED UPON EXCAVATION
- EVEN SUBTLE FEATURES MAY HAVE OVERWHELMING EFFECT

Figure 3. The material parameters: Rock-soil-water mixed ground.

the material that is at the site, we need to explore it and observe it in order to evaluate it.

In many situations, we may be able to avoid problems, so siting can be very important. Returning to the previous figure, you would find a situation that would be much improved if you could stay down in the clay and never pop up into the overlying sand. An adjustment in vertical alignment can be of substantial benefit to a project, and much can be gained by having a project which is fitted to the ground to take maximum advantage of the site geology.

It is difficult to sample and observe the material before tunneling through it. The view from the bottom of a tunnel or from the end of a borehole is often poor. When you are pumping grout and do not know where it is going, you can almost envy NASA's space program. It seems so easy for them to communicate and to observe in space. But underground space is opaque to many of our sensors. We may utilize remote sensing methods, in addition to drilling holes. But all these methods have limitations and are difficult to interpret in such a way that we can predict how the ground is going to affect the tunneling process. The feeling that we are poking about in the dark, missing more than we hit, is certainly justified.

One of the sources of information that is used in an exploration program is observation of surface exposures. Here (see Fig. 5) we are viewing an exposure of rock, 2,000 feet above a proposed tunnel. This happens to be the Stillwater Tunnel that Ken Schoeman was describing in his presentation earlier this morning. The slide shows closely spaced joints and fractures in a limestone, but we are 2,000 feet above the tunnel. The tunnel is in another material. This is limestone and the tunnel is in shale. Furthermore, the stress levels will be much different down in the tunnel, and the rock may be less weathered at depth than at the surface. It is obvious that you are making quite an extrapolation to try to evaluate what will be encountered down in the tunnel from surface

	PORTION OF ROCK MASS CONTROLLING BEHAVIOR			
ENVIRONMENTS	(1) Intact	(2) Joints, Bedding	(3) Shear Planes or Closely-Spaced Joints	(4) Soil-like Zones: Sheared, Broken or Decomposed
(A) Body Forces (Loosening)				
(B) Insitu Stress: Shear Stress Effects 1) Stress Slabbing				
2) Squeezing				
(C) Insitu Stress: Volumetric Effects 1) Swelling				
2) Slaking				
(D) Water Volume of Inflow, Erosion, Piping, Solution, Pore Pressure				
(E) Gas Explosions, Toxicity, Pressure				
(F) Temperature Effect on: Rock, Lining, Personnel, and Operations				

Figure 4. Rock failure mechanisms in underground openings.

information. On the other hand, drilling to depths of 2000 feet is expensive and difficult. Furthermore, drilling gives a very local picture of the conditions and does not tell you the extent of major features such as shears, faults and shear zones. One must try to integrate all available information, recognizing the limitations of each method.

Another example of sampling procedures is illustrated by a storm water tunnel project in Phoenix. Much of the soil is alluvium, containing a very heavy concentration of cobbles in a matrix of sand and gravel. Small-diameter borings were drilled, but the information obtained from small diameter samplers and borings in regard to the characteristics of boulders and cobbles which are much larger than the hole diameter is not very informative. One of the very interesting things that was done on the Phoenix project was to drill four large-diameter holes at the time of the bid, when all the prospective bidders were at the site. A protective casing was installed and then at least 50 individuals were lowered into each of these bore holes to observe and visualize the ground conditions for themselves: they could actually see the boulders. The use of something other than just a standard 2- or 3-inch-diameter sampler can be very helpful in many situations.

It is difficult not only to predict the geology and the ground conditions, but one of the major problems is being able to predict performance (see Figure 3). In many cases when difficulties are encountered, it is not so much that there is a change in the geology, but there is a change in our perception of how the geologic materials behave, how they affect the construction of the tunnel. The column on the left of Figure 4 illustrates the ground conditions, or environments, in which a tunnel can be driven. Several stress conditions are shown. First, there are the body forces, or the effect of loosening and movement resulting from the dead weight of the material surrounding the tunnel. Second, there is the effect of shear stresses tending to cause the ground either to fracture brittlely or to squeeze or creep into the opening ductilely, and third, the volumetric stresses which cause swelling or slaking of the material. Other environmental conditions include water, gas, and temperature. For any of these environments, it is necessary to define indices, obtainable from simple, reproducible tests that can be correlated with the observed movements and failure in a tunnel. For example, what is squeezing and how is it occurring? Is the squeezing taking place primarily in the intact portion of the rock mass? Is it taking place along individual joint surfaces? Is it occurring in the broken, sheared rock in fault zones? And to what degree is it occurring? If we can carry out simple index tests that will give us that information, it would be quite helpful. Once we have devised an index test, we must relate it to the actual performance of the tunnel. Correlations of actual field performance are coupled with the index properties by the use of appropriate mechanical models. To do this requires an understanding of the theory of the mechanics of the behavior as well as the practical aspects of tunnel construction and a keen eye that has observed what rock, soil, and water do when mixed with a tunnel. At that point we start to obtain a much better understanding of the behavior underground. Rather than taking some index and saying, "because the joints are spaced 2 feet on center that means I should install 10-foot-long rock bolts 5 feet on center," we should look more closely at just how the rock indices, such as joint spacing or compressive strength of a sample, affect the actual behavior in the tunnel. Will the rock slab? How deep are the slabs? Once we understand these failure or behavior mechanisms, we can do a better job of selecting support and evaluating the potential progress of a tunnel.

Figures 5 and 6 illustrate conditions in the Stillwater Tunnel, and the environments and ground conditions encountered. We found that there was some squeezing in the tunnel in jointed and faulted zones. Raveling and loosening along joints and shears also took place and tended to cause more difficulties, in many respects, than the squeezing. Raveling affected the ability to drill holes up into the crown when little pieces were coming out the bolt hole could not be held open. It affected grouting of the concrete segments, as I will discuss later.

The stress slabbing and fresh fractures that formed in the tunnel were relatively minor. (The ratio that Ken Schoeman was discussing--unconfined strength divided by the <u>in situ</u> natural stress--can be used to evaluate stress slabbing characteristics. When that ratio is of the order of five or less, minor stress slabbing and fracturing occurs around the tunnel. The squeezing condition is not defined by that ratio.)

At Stillwater Tunnel, the progress was not significantly slowed by the squeezing conditions. Squeezing resulted in several inches of inward movement and some bowing of the lagging in the steel rib section of the tunnel. The tunnel boring machine advanced through the steel rib sections, even when squeezing was occurring, at an average daily rate of approximately 150 feet a day over a period of several months. So squeezing did not have a major influence on the progress of the tunnel. The conditions that did affect support and progress were more subtle. Raveling in the arch tended to cause difficulty when grouting to fill the annulus between the rock and the concrete segments which were placed behind one of the TBM's (see Fig. 7). It is very difficult to develop a seal and prevent grout from leaking out of the space between lining and the rock in a situation where the ground has raveled and left a void in the crown or has already moved down on top of the tunnel. In sections where raveling did not take place and the rock surface was smooth and circular, as we see in the lower part of the diagram of Figure 7, movement of the lining tended to be all inward. In situations where the rock had loosened at the crown, the movement was inward in the sides but it was upward into the crown of the tunnel as the lining displaced vertically into the void or loosened zone caused by raveling (upper part of Fig. 7). That situation caused distortions and bending movements to develop in the lining, with minor cracking of concrete segments. The combination of, first, the raveling which prevented tight contact between the support and the rock, and, second, the squeezing effect of the ground caused inward movement that resulted in distortions which, in some cases, could crack segments. So, raveling influenced the behavior strongly.

Slaking was another condition encountered in the tunnel. Small changes in humidity can affect the way the rock breaks up, and is often very difficult to predict. But there are ways one can test these materials. If samples can be obtained, controlled humidity tests are quite useful. They are beginning to be correlated with actual field experience so that slaking conditions in other tunnels can be predicted.

In concluding the discussion of the constraints on tunneling resulting from the characteristics of the material, we should note that not only the material characteristics but their effect on the performance of the tunnel are difficult to predict. Use of index tests, combining them with an understanding of the mechanisms of failure, and the behavior of the tunnel in terms of the construction, the geometry of the tunnel--relating it to the material itself, the rock and the loadings--putting all that together, then I think you begin to get a picture as to how the material properties can be related to the actual progress of the tunnel to help us in our prediction efforts.

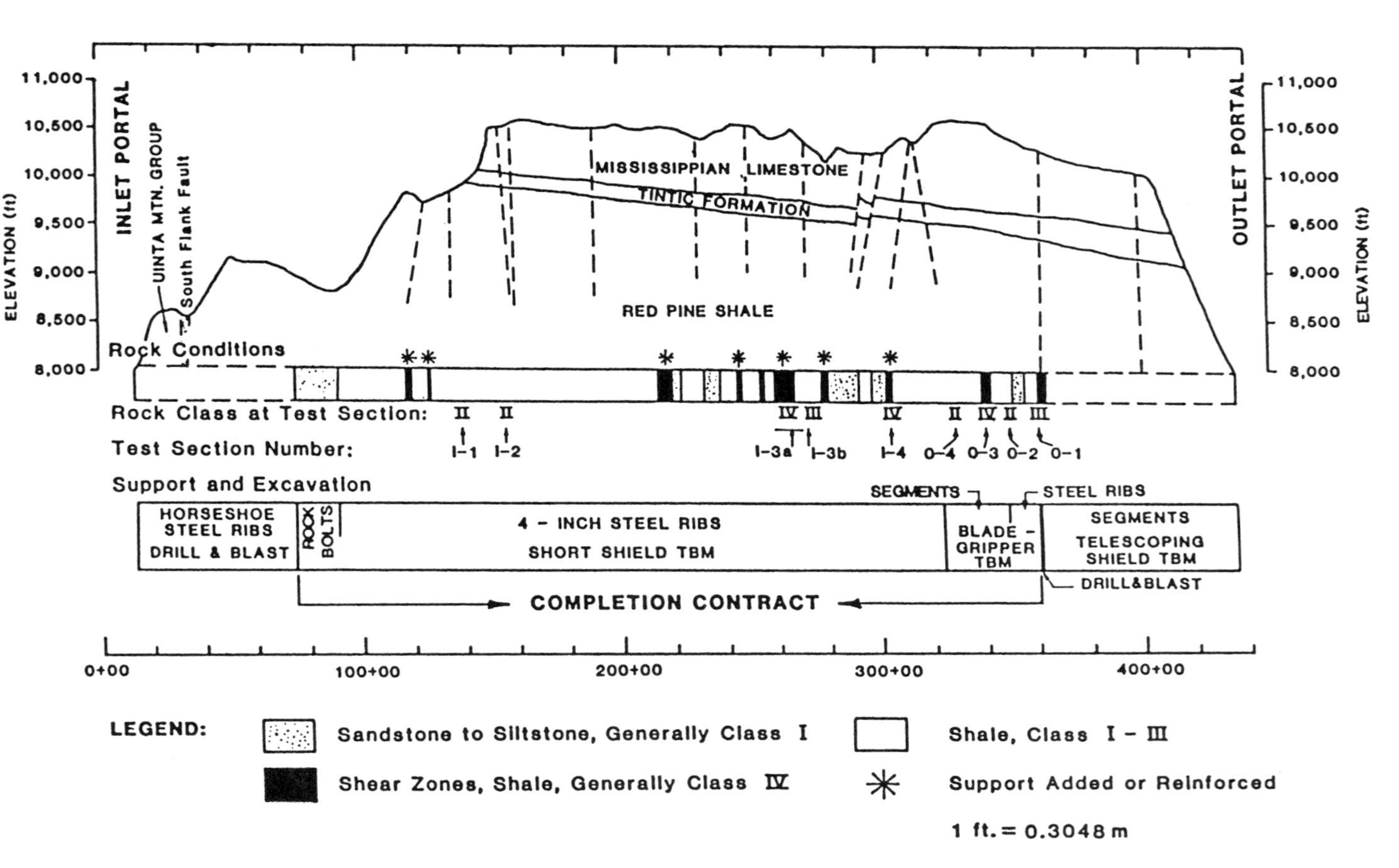

Figure 5. The Stillwater Tunnel.

ENVIRONMENTS	PORTION OF ROCK MASS CONTROLLING BEHAVIOR			
	① Intact	② Joints, Bedding	③ Shear Planes or Closely-Spaced Joints	④ Soil-like Zones: Sheared, Broken or Decomposed
Ⓐ Body Forces (Loosening)	SOME LOOSENING OF STRESS-FORMED SLABS	←	RAVELLING OF SMALL PRISMATIC PIECES	→
Ⓑ Insitu Stress: Shear Stress Effects 1) Stress Slabbing	FORMATION OF THIN SLABS, IN FRESH ROCK OR ALONG INCIPIENT FRACTURES			
2) Squeezing			CL III: 2-3 IN. DIAMETRAL CLOSURE WITH TIME	CL IV 4-6 IN. DIAMETRAL CLOSURE WITH TIME
Ⓒ Insitu Stress: Volumetric Effects 1) Swelling			SOME INVERT HEAVE WHERE WATER IS PRESENT CL III	CL IV
2) Slaking	MINOR TIME-DEPENDENT SLABBING; HUMIDITY DROPS TO 90%	—		
Ⓓ Water Volume of Inflow, Erosion, Piping, Solution, Pore Pressure	No significant water problems			
Ⓔ Gas Explosions, Toxicity, Pressure				
Ⓕ Temperature Effect on: Rock, Lining, Personnel, and Operations				

Figure 6. The Stillwater Tunnel: Rock failure mechanisms in underground openings.

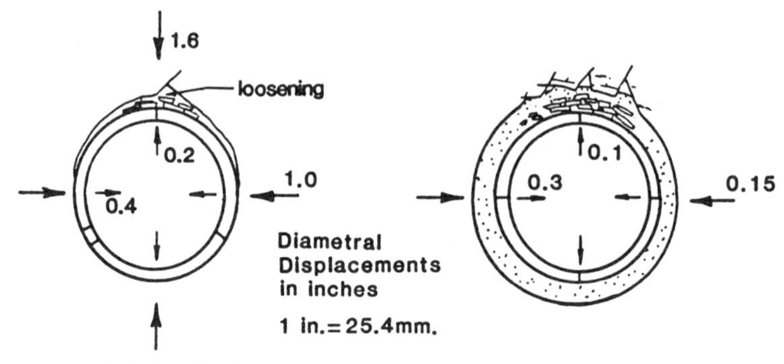

4 - INCH STEEL RIBS
TEST SECTION 1, OUTLET
CLASS III

5 - INCH CONCRETE SEGMENTS
TEST SECTION 4, OUTLET
CLASS II

Outward deflection of the lining due to loosening in the crown.

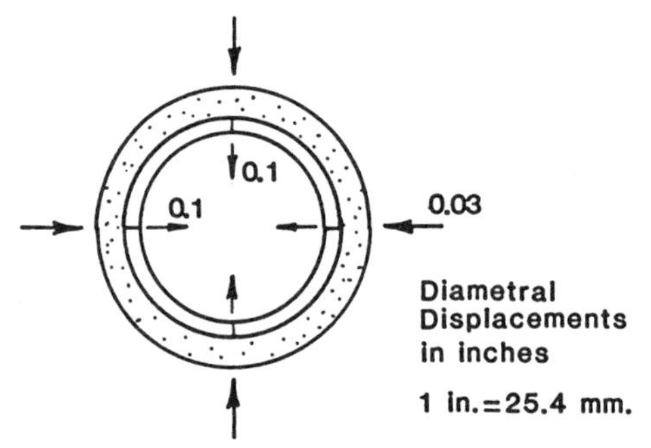

5 - INCH CONCRETE SEGMENTS
TEST SECTION 2, OUTLET
CLASS II

Inward movements of the lining.

Figure 7. The inward/outward movements of the lining.

Often we only obtain information about a project after the excavation has taken place. Prior to construction of a deep tunnel in Colorado, the Henderson Haulage Tunnel, driven through the Continental Divide, we inspected the regional geologic map, and found that there were many fault zones throughout the Front Range, but none in the vicinity of the proposed tunnel. Was it because there were no fault zones at the proposed site? Or perhaps, the data suggested that fault zones have an affinity for tunnels because in the areas, where faults were present, tunnels had already been driven. Of course, the reason for the presence of faults was that many of the fault zones had been discovered by tunneling into them.

Even subtle features can have very large effects on a tunnel's progress. For example, if a sand has a very slight amount of cohesion, it may stand unsupported so that it will not run into a tunnel. That is the type of condition that is difficult to evaluate from a sample of sand, which is usually disturbed in the process of sampling so that it loses its

- ALREADY IN PLACE

  MAINTAIN STABILITY
  MINIMIZE MOVEMENTS
  INITIAL CONDITION CRITICAL

- HIGH

- UNLIMITED BUTTRESS

- REDUNDANT

- "THE ADVANTAGE OF BEING A BEAM"?

Figure 8. Constraints on tunneling resulting from the load.

cohesion. Very slight changes in the characteristics of the material can make a very large difference as to how the ground will perform.

Figure 8 outlines the constraints on tunneling resulting from the load, the second main heading of our outline. Of greatest importance is the fact that the load in a tunnel is already in place: the tunnel must be supported and advanced with the load in place. It would be desirable to be able to place the support before we have to excavate the tunnel. We cannot always do that, do not always need to do that. But in some situations, tunneling techniques are designed to presupport the ground so that it is safe to operate in the front end of the tunnel working under the load of rock or soil that is already in place.

The primary problems in tunneling arise at the tunnel heading, e.g., the problem of maintaining the stability of the tunnel so that it does not collapse, and very practically and perhaps very predominantly, the problem of providing safety for the people working in the tunnel face. Traditionally, the tunnel heading has always been under the control of the crew that is in the front end of the tunnel. And I think this is one of the factors in tunneling that makes it difficult for the designer to get a good handle on what is going on in the project or to specify what will be done, because the person at the heading is the one that is at risk and the one that is placing support to provide for his own safety. The designer is traditionally responsible for the permanent lining and may require that a final support be placed that will satisfy the long-term requirements, regardless of the initial support placed by the contractor. The more that initial support and final support can be integrated and put together, perhaps as one support, or at least be made compatible, then we are making progress. For example, precast concrete segments are increasingly used as initial and final support on Metro tunnels in soil. But not always is the initial support going to work as the final support, and that should not be our goal.

The other situation we have in the front end of a tunnel is the problem of minimizing ground movements (Fig. 8). Even if stability can be maintained, we may have a situation in which there is a potential for enough ground settlement to result in damage to utilities and structures.

Spiling out over the top of the tunnel is one way of providing support ahead of the tunnel face. The spiling is placed under one steel rib and over the steel rib that is closest to the front end of the heading, in order to support the ground several feet in front of the tunnel so that you can then go in and excavate out a little bit more under the umbrella protection of the spiling. This is a classic soft-ground tunneling technique, but it has been used effectively in some situations with boring machines in rock tunnels where weak rock and soil were collapsing onto the machine cutterhead. Ground movements in the front end of the tunnel are occurring because the load is already in place. Shields are used in many tunnels to provide initial support and control ground movements that could result in damaging settlements.

Movements cannot only take place in the front end of the tunnel but around the tunnel shield as it is advanced, and at the tail of the shield as the lining is installed. An example is the Washington, D.C. Metro. Concrete segments are placed in the tail of a Lovett Wheel-type excavator machine. After the segment is placed, and as the shield moves forward, grouting is carried out to fill the annular space between the shield and the lining before the freshly exposed ground can collapse into the 2- to 4-inch annulus. The speed, efficiency, and completeness of grouting are going to have a lot to do with how much settlement takes place at the ground surface.

There are other procedures for controlling ground conditions around the tunnel. Figure 9 illustrates one method that has recently been used on several projects in lieu of underpinning. It was developed on the Bolton Hill project in Baltimore, by a specialty contractor in grouting, the Hayward Baker Company. The idea here was to eliminate the loss of ground by pumping very low slump compaction ground from the ground surface as the tunnel was advanced beneath the grout pipe location. The compaction grout would first densify the loosened soil and fill the voids that existed around the tunnel and then, with further application of the grout, begin heave of the soil and even recover movements that had occurred previously. We used the system in Minneapolis, where tunneling was taking place 15 feet below an existing old stone arch culvert tunnel that had been built near the turn of the century and was flowing with water at the time the tunnel passed beneath it. By placing a communication system between the tunnel, the culvert, and the surface and by making measurements of settlements in the old culvert, we were able to determine when to start and stop the placement of grout so that the culvert would neither settle nor heave excessively (Figs. 10, 11 and 12). If the base slab of the culvert had been cracked, then the water flowing in the culvert could have flooded the tunnel being constructed. Techniques of handling ground conditions like this are being used more frequently, and nowadays there is much more understanding of the ground conditions and ways of controlling ground in urban tunneling.

Continuing with the outline of Figure 8, we note that the weight of soil or rock above a tunnel is very large, and loads can be very high but, in our favor, the rock and soil surrounding the tunnel also serves as an unlimited buttress. In many cases, the rock or soil itself is the structure, and can support itself with minimal, but appropriate, assistance from support elements placed into or against the ground by the tunneler.

Structures located above the ground surface perform much differently. Gordon describes this well in his book, Structures or Why Things Don't Fall Down. He described "the advantage of being a beam." And, of course, if you have an above-ground structure whose roof is a beam, there is no outward thrust to push the walls out, so there is an advantage to being a

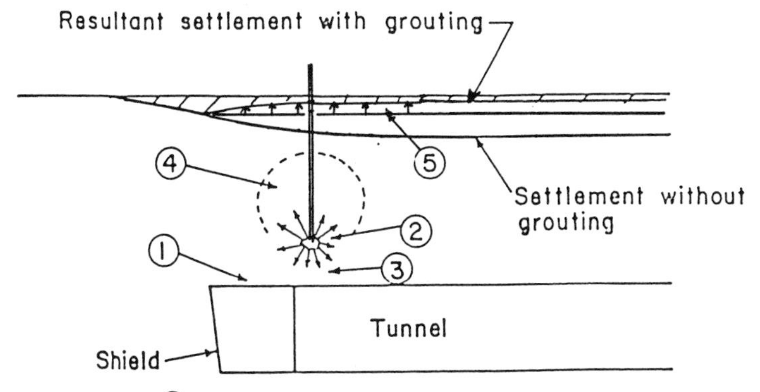

① Ground loss over front of shield.

② Insertion of low slump compaction grout behind tail of shield.

③ Densification of soil around grout bulb, recovery of volume loss in vicinity of tunnel before it can reach the surface.

④ Heave of soil above tunnel.

⑤ Recovery of previous surface settlement.

Figure 9. The effect of compaction grouting on surface settlement.

Figure 10. The control of compaction grouting during tunneling, Minneapolis.

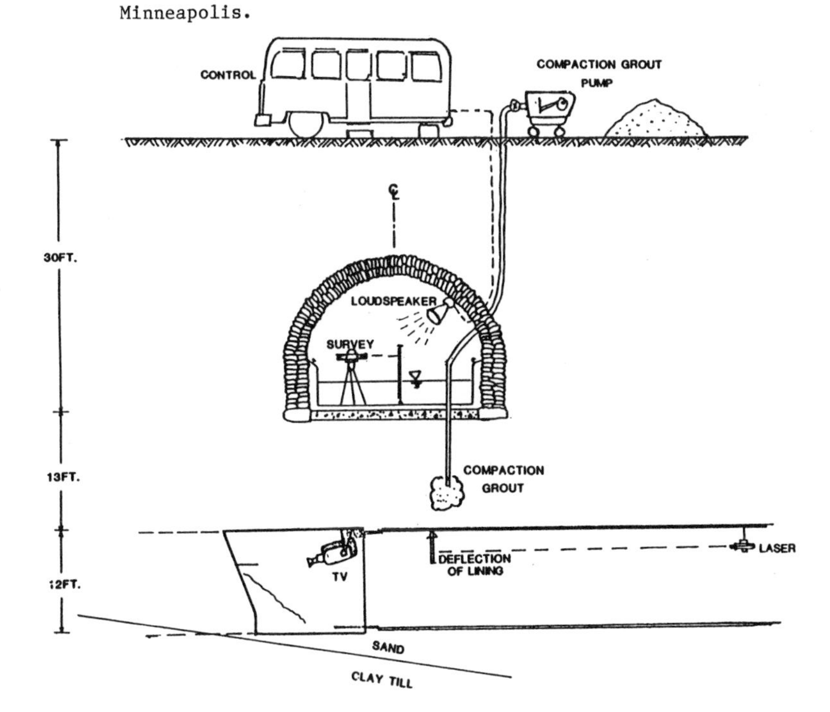

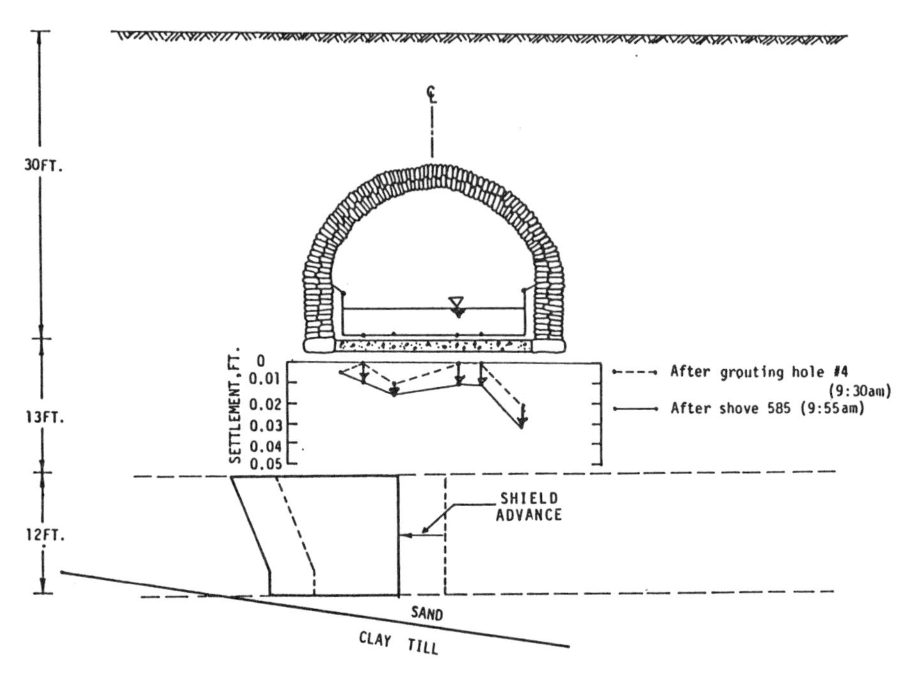

Figure 11. The settlement of culvert due to one shove of the shield.

Figure 12. Heave of the culvert due to compaction grouting through hole 5.

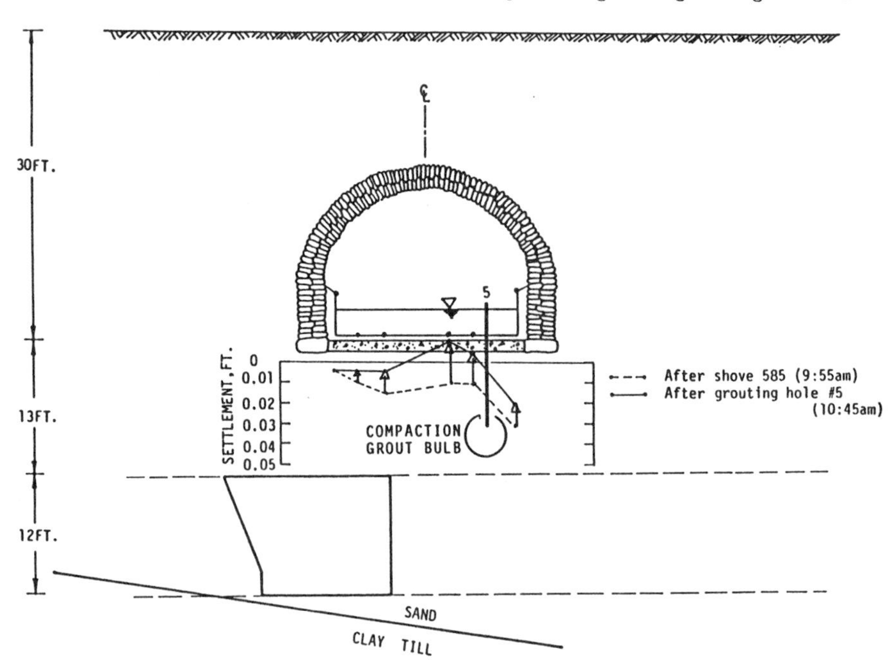

beam. In some of the cathedrals in Europe, the outward thrusts from the arch resulted in instability of the walls. In response, flying buttresses were placed, and statuary was added on top of the buttress to provide the vertical force that would keep the thrust line in the buttress and prevent collapse (Fig. 13). Similar behavior occurs in a tunnel. The reaction from the soil or the rock against the lining will keep the thrust in the section of the lining (Fig. 14).

Furthermore as the lining deflects outward when it is loaded, the side reaction increases, tending to maintain stability. The lining remains stable even though we apply very high loads. We find that, if we build the tunnel as an arch, using the soil or rock as a reaction, that only a relatively light lining is required. We also find that a concrete tunnel lining does not require the heavy tensile steel reinforcement that would be calculated from the American Concrete Institute Code for above-ground beam-columns, because the reaction provided by the ground keeps the concrete lining stable, even though local tensions develop in the concrete.

Now returning to the Stillwater Tunnel, the steel lining in this tunnel could be very tightly placed and expanded against the ground surface, which was cut to a smooth radius by the tunnel boring machine. It was able to essentially become fully loaded to its yield capacity (fully plastic) because it had very good contact with the surrounding ground. When the lining started to kick out longitudinally along the line of the tunnel, some collar braces and extra reinforcement were placed in that direction to help the lining to maintain its position and continue to act to provide support, even though it had reached its maximum capacity. Had this steel not been restrained by the ground, it would have collapsed.

Trying to design a structure underground as if it were a beam results in a very, very heavy support, extremely large sections (Fig. 15). Alternatively, an arch is placed and the result is a much better configuration.

I wanted to make just one or two comments here about geometry, the last constraint listed in our outline (Fig. 1). Of course, many tunnel projects are long and linear. In a tunnel, everything is happening at the face and progress is not obtained unless the face advanced. We then are concerned with the actions of a few individuals at the tunnel heading or a machine that may occupy the entire heading. Here is a situation in one of the deep water supply tunnels on the Central Utah Project. A tunnel boring machine was being advanced into a faulted sandstone so the fault consisted largely of sand (see lower portion of Fig. 16). The support and excavation methods were not typical of TBM tunnels. What was done in this situation was to combine some techniques which were quite old and had been used for years, similar to what was done on the Moffatt Tunnel, which was built in the 1920s (see Fig. 17). The TBM was stopped and the miners began hand mining ahead of the machine. Soft-ground tunneling procedures were used to open up small headings at the top of the tunnel, supporting each with steel and timber, then finally combining the small tunnels into one larger supported arch that provided the overhead support that permitted the TBM to be "walked through" ground it could not have mined through. The contractor was able to do this because he had a boring machine with an open cutter head that allowed the men to go through the front end of the machine. He had a short shield on the machine so that he could get close to the front of the tunnel and place support up over and in front of the cutterhead. And he had tunnelers experienced in a variety of tunneling techniques and ground conditions. He was able to move rapidly through the good ground, and make slow, but steady progress through the difficult ground.

Figure 13. Flying buttresses in European cathedrals.

Figure 14. Thrust in the lining section.

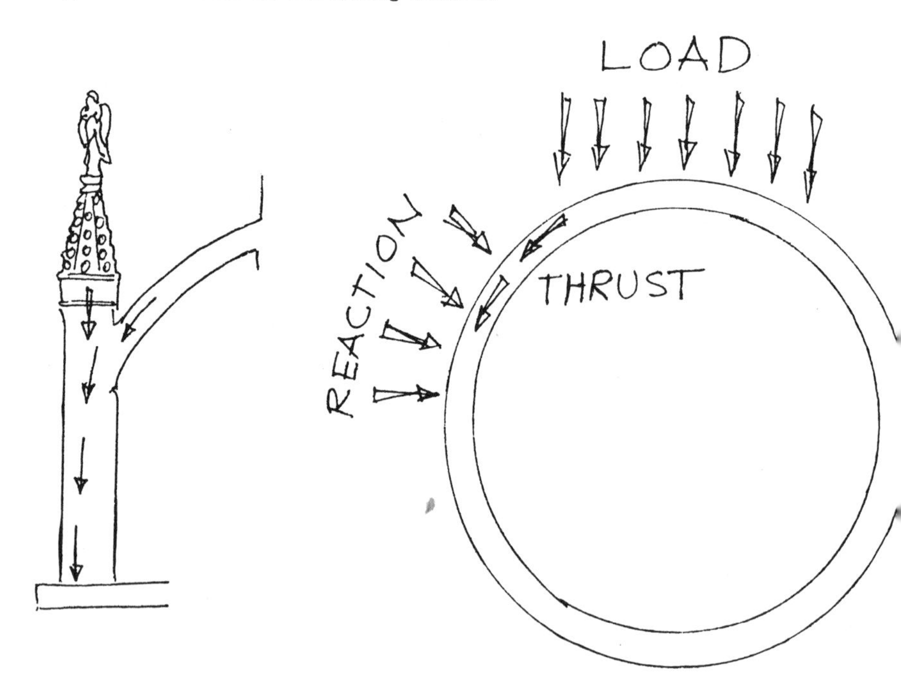

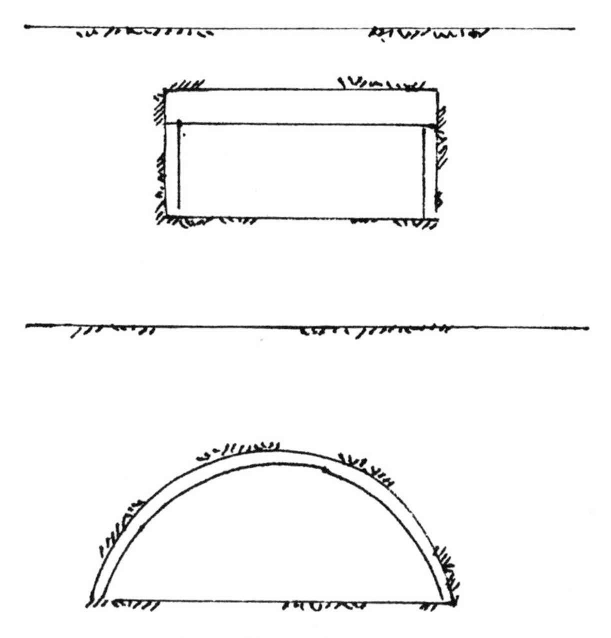

Figure 15. Beam versus arch configuration.

Figure 16. Tunnel boring machine being advanced into faulted sandstone.

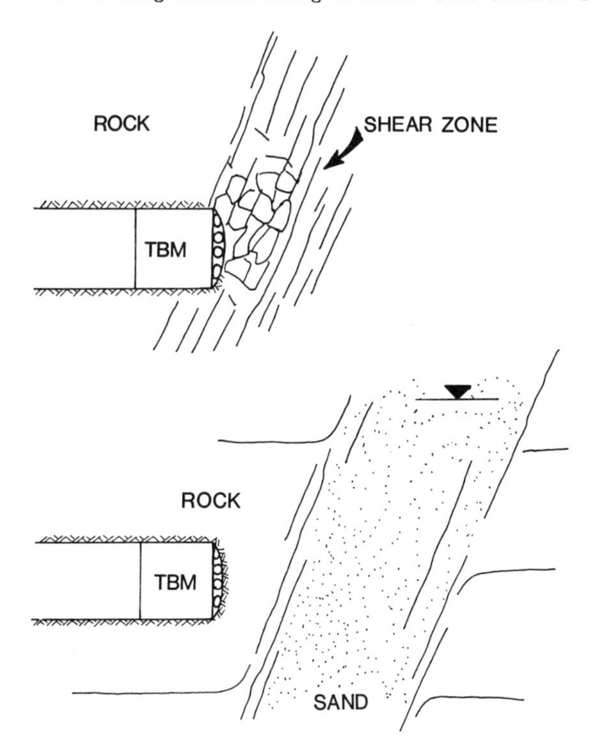

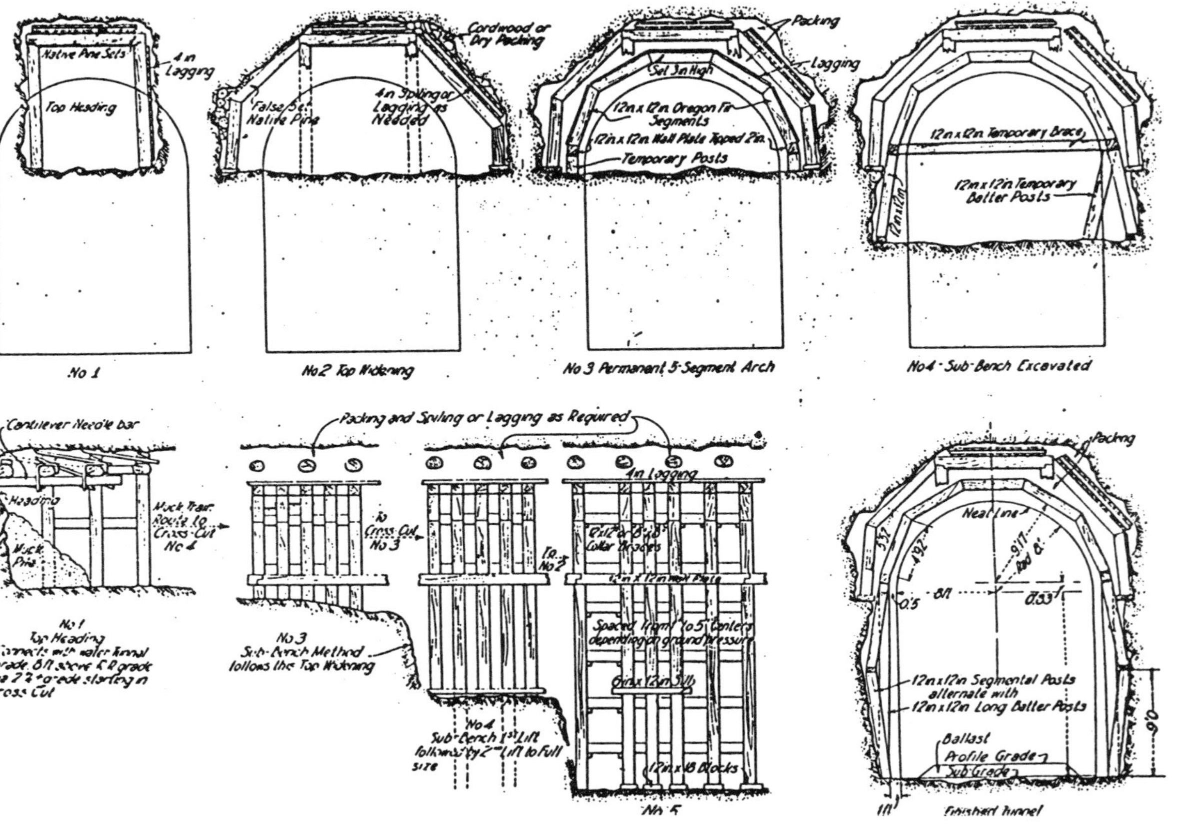

Figure 17. Traditional support and excavation methods used with Moffatt Tunnel

I think we recognize that tunnel boring machines are capable of very high advance rates; they have high capital costs, and there is also the risk of large delays (Fig. 18). They are sensitive to certain ground conditions, and therefore it is very important that we be able to predict and evaluate the range of conditions that may exist on a given project. We need a tunneling system that will be able to make good progress in most of the ground and be able to move at least consistently, although perhaps slowly, in the more difficult type of ground. We need to have systems in which tunneling, excavation, and support methods are well integrated, so that the operation is efficient and capable of coping with the ground conditions that are encountered.

One of the challenges at the Stillwater Tunnel, was to design a tunnel boring machine that would be able to advance through squeezing ground and install concrete segments. Those requirements controlled, to a large extent, the characteristics of the tunnel boring machine that was used both on the first and the second Stillwater Tunnel contracts. The tunnel boring machine design and the lining design had to be compatible. On the first contract, the shield placed around the machine to permit installation of the concrete segments could not be effectively advanced through the squeezing, raveling ground. The redesigned blade-gripper shield on the second Stillwater contract was capable of advancing through the same ground.

One of the items I would like to emphasize, shown at the bottom of Figure 18, is that we really need to have a contractor working with the designer. In most of our contracts, the contractor looks at the project only after it is designed. How is he going to approach that bid as he

Figure 18. Tunnel-boring-machine factors.

TUNNEL BORING MACHINES

- CAPABLE OF HIGH ADVANCE RATES
- HIGH CAPITAL COST
- RISK OF LARGE DELAYS
- SENSITIVE TO CERTAIN GROUND CONDITIONS

INTEGRATION OF TUNNEL EXCAVATION & SUPPORT METHODS. . .

Essential.

INITIAL & FINAL SUPPORT. . .

Should be coherent plan.

SEPARATION OF DESIGN & SPECIFICATIONS FROM CONTRACTOR'S PLAN. . .

Creates Incompatibilities.

looks at the job if he can see that there is a much better way? Not only that, but he may recognize that if the contract is carried out the way it is specified, there is a very great likelihood of major difficulties in construction of the tunnel. How can the contractor's expertise be brought into the project at an early stage so that the project can benefit up front from his innovation, and experience? I think that the example Ken Schoeman was describing for Stillwater Tunnel is one that is good, because, for the second Stillwater contract, the Bureau of Reclamation negotiated a contract and made their selection of a contractor on the basis of both technical merit and cost. Traylor Brothers/Fruin Colnon was the contractor selected. The contractor was able to develop a new machine, the blade gripper or walking shoe TBM that Ken Schoeman described, and I do not think that that would have been possible if they had a strictly low bid, hard money, and fixed-cost contract. The blade-gripper TBM that resulted will have future use on other projects.

As summarized in Figure 19, we need to recognize the fact that the ground is variable and it's difficult to predict, and the universal tunneling machine has not arrived. We cannot handle all ground conditions with a given machine, and in some cases, even on a given project, the machine selected may have difficulty in some of the ground. So there is a risk. There is a need for flexibility in the way we contract and the way we carry out the work, and I think that is one of the major limitations of our tunneling technology. We may find that we have a design that may not fit the ground or the contractor's plan. It is very important that we be able to adjust the design as required, or that the contractors be able to work with the designer to achieve the best product. We very often find that contracts lack flexibility. The incentives are not there for the contractor to work together with the engineer in completing the job. The contracts do not encourage engineer/contractor teamwork.

Figure 19

GROUND IS VARIABLE-DIFFICULT TO PREDICT

UNIVERSAL TUNNELING MACHINE HAS NOT ARRIVED

DESIGN MAY NOT FIT GROUND OR CONTRACTOR'S PLAN

CONTRACTS LACK FLEXIBILITY, INCENTIVES: DO NOT ENCOURAGE ENGINEER-CONTRACTOR TEAM.

MECHANISMS TO ENCOURAGE INNOVATION, RESEARCH, AND DEVELOPMENT ARE LACKING.

There is another condition in tunneling that we do not see enough of: a mechanism to encourage innovation, research, and development on a project. Who will support research and development? Where is it going to occur? A small contractor cannot do it; he does not have the assurance of continuing participation for example on the major project. He does not know if he will get the next job. He is really designing his equipment for that specific project. Manufacturers are not large. We do not have General Motors or IBM type corporations in this industry. The owners, I think, have some opportunity here because they may have a major, continuing interest in a project or in a region. The proposed Superconducting Supercollidor project (where there may be on the order of 100 miles of tunnel) presents a unique opportunity for tunneling research and development that can benefit the project, and also the industry. It would also benefit greatly, as would many of our tunneling projects, from the use of contracting practices that would encourage contractor innovations, and provide incentives for new developments and efficiencies on the project.

What Is Ahead in Mechanized Tunnel Excavation

RICHARD J. ROBBINS
President
The Robbins Company
Kent, WA 98031-0427

Yesterday afternoon as I was traveling to Boston from London, I read in the Financial Times an article about a project in Norway. They have a special problem there which fits in well with this conference on macro-engineering and underground excavation. Two of the chief medical researchers and examiners involved in deep underwater medicine for Norway have resigned in protest over that country's plan to set up offshore oil platforms in 300-meter-deep water, which requires the divers to actually be working at the bottom in those depths. They claim that the authorities aren't heeding their warnings that that is going to be a dangerous situation for the divers. In the same article, they point out that a 55-kilometer tunnel is under consideration that will go out offshore and be used as a collection and production facility beneath the seabed so that gas and oil can be collected and pumped back to shore instead of using these large platforms. That is a 34-mile tunnel, and the interesting thing about it, according to the article, is that they expect a cost saving of 6.5 billion Norwegian Crowns; that is about $750 million, compared to using an offshore platform for the work. That puts it in the class of a macro-project when you consider the savings they are talking about.

The article does not describe the job, but the tunnel for this work will start on an island in igneous rock. There will be shafts and declines, and then a long tunnel going out through a series of sedimentary rocks which will go through a whole series of faults, shear zones, some very weak rock, some that they expect to flow under the pressures that they have to tunnel through, and through a gaseous condition. You can imagine there is a lot of potential in a project of that type. That takes real long-range planning.

THE FUTURE OF MECHANIZED TUNNELING

Now, as to the future, my talk is going to be about what to expect in mechanized underground excavation. Looking at the future, due to the recessions in both heavy construction and mining, the size of the market for underground excavation in general has shrunk to the point where a lot of money that was being spent four or five years ago has dried up, that is, research and development money, both government and private. This is definitely having an effect on what will be developed in the future. However, a number of projects have been started and are underway, and some of those will be completed.

For the purpose of describing what to expect in mechanized excavation, I have broken geologic conditions into two classes: competent rock tunnels and difficult geologic conditions.

COMPETENT ROCK CONDITIONS

Speaking first about competent rock, I expect that most improvement will be of an incremental and developmental nature, that is, there probably will not be major steps. We seem to be up against a development barrier; we may have gone about as far as we can go without some major steps. There are some important potentials for development, but most of

Tunneling and Underground Transport: Future Developments in Technology, Economics, and Policy
F.P. Davidson, Editor

those are related to materials technology: better steels, particularly for the cutters, steels which will have high impact resistance and also high strength. This may require composite materials. But developments will probably be incremental.

One of the things that we have to do is at least perform at the level of the state of the art, which is something we seldom do. The first problem is to recognize what the state of the art is. One of the difficulties we get into is that we seem to "reinvent the wheel" all the time. We are solving problems that have been solved before, either by other members of our own organization in past years or frequently others that we do not know about, and then in addition to that we make some changes which are frequently regressive. We end up with a mixed bag of some brilliant new ideas, some things which are really state of the art, and some things that are not as well done as they were some years before. The other problem is that we frequently discard ideas that were developed by other organizations due to the "not invented here" syndrome.

WATER-JET ASSISTED ROCK CUTTING

We can expect some advances in rock-cutting technology. One of the things under investigation all around the world at this point is water-jet assisted cutting. We went through a period where technology was heavily funded by various government agencies to develop water jets at very high pressure levels and high-velocity cutting. We are now investigating much lower pressures, and rather than cutting the rock, it is basically chip removal. That is being investigated now in the United States by several organizations and in South Africa, England, and Germany. I would not be surprised if it is also being investigated in other areas.

Figure 1 shows a 7-foot tunnel boring machine with some high-pressure water jets shooting out of the front. You can see the placement of the jets and the disc cutters near the center of the machine. That is granitic rock in which this test was made, and you can actually see grooves, some of which were cut by disc cutters and some by the water jets. At this time water-jet assisted cutting is being carried on probably in the 2000-5000 lb/in^2 pressure ranges. The work that was done when these photographs were made was at 10 times those pressures, up to 60,000 lb/in^2. What we expect to achieve at the lower pressures is improved lubricity, reducing the torque required on the cutter head. If you maintain the same power, you can thrust the machine at higher levels without running out of torque or horsepower. This will result in higher penetration rates. We may also be able to reduce dust generation at the same time.

NONCIRCULAR HARD ROCK TUNNELING

Another feature that we expect to see happening in good geologic conditions is the development of noncircular tunnels for highways and railroads. In Bergen, Norway, the Norwegian Road Department is building a pair of highway tunnels with a circular tunnel boring machine 7.8 meters in diameter. That is nearly 26 feet in diameter. The tunnels are going through granite. The problem is that the size of that bore is not big enough for a two-lane highway tunnel unless you cut out the bottom corners. They are boring the tunnels with a machine, then coming through later and blasting out the corners. Figure 2 shows a view of the tunnel on the right as it would be if blasted with a concrete lining. In the middle, it shows the way they intended to construct the tunnels with the 7.8-meter machine with blasted corners to make the roadway wide enough, and on the left it shows a system of widening that tunnel mechanically. This can be done with a machine which either follows the tunneling

Figure 1

Figure 2

machine or works right together with it by boring simultaneously with the machine with big drums to enlarge the tunnel in that shape.

Next we shall consider the Rogers Pass Railroad Tunnel. We have heard this job described in previous chapters. The contractor has chosen to use a tunneling machine, in this case 22 feet in diameter, or 6.7 meters in diameter, to bore the circular part of the tunnel. What they now plan to do is bore the upper part and blast out the lower part and then line it with concrete, making that center section. But, in the meantime, they have decided it may be feasible to do what can be done to mechanically excavate the invert with a series of big drums with cutters on them, and we believe that this is feasible. If work of this type comes up in the future in significant amounts, I believe that special excavation machinery will be built to cut these noncircular tunnels so that we can make highway tunnels and railroad tunnels, out of circular tunnels which is a shape that does not really suit the highway or railroad engineer.

NONCIRCULAR TUNNELS FOR MINING

The next topic is noncircular tunnels for mining, and to give you an idea, I include Figure 3. It is a prototype tunneling machine for use in mining, mounted on crawler tractors. This machine is operating at Mt. Isa Mine in Australia in a hard quartzite rock. It uses disc cutters and the boom swings from left to right, cutting a noncircular shape. We first tested this machine in a granite test site near Seattle. It is now possible to make a noncircular tunnel in hard rock, which is something that heretofore has not been possible. The roadheaders, which were used only to mechanically excavate a noncircular tunnel before, are not capable of doing that job in this type of rock.

DIFFICULT GEOLOGIC CONDITIONS

On difficult ground conditions, squeezing ground is, of course, an area in which we will see future development. The U.S. Bureau of Reclamation job at Stillwater in Utah is an example of such a job. A double shield machine which was trapped by the squeezing ground was modified in the tunnel to use a front walking gripper mechanism. I believe that system will be developed for future jobs.

The other thing that I think you will see in bad ground is the combination of features for hard-rock boring and soft-ground boring such as the slurry shields or earth pressure balance shields. The features of all these machines have been developed independently and proven, and there have been a couple of tentative attempts to marry these features. I believe that a machine can be built that is relatively insensitive to ground conditions. A machine is needed that can bore through soft ground, enter very hard rock, convert its muck handling system, and continue to operate in that sense in a mixed face condition and then go back into soft ground.

SHAFT BORING MACHINE

Shaft boring machines have been used for blind shaft drilling in Germany. The muck is pumped out of a shaft. The advantage that you are looking for here is basically speed. The machine was used by the Cementation Company in the United States on an experimental basis using an impact breaker to break the rock and a backhoe to pick it up. Shafts are sunk upside down or vertical boring--an Air Force tunnel borer, pointed straight up. A new shaft sinking machine has been completed which will mine the rock with a disc cutter head and move the rock out with a clam

Figure 3

proceeding ahead of a bulldozer blade. This machine has yet to be proven, but is now built and will be put into service fairly soon.

The water we expect to use will probably not be much more than two or three times the water that is now being used in conventional dust suppression systems. And one of the things which gives us some hesitation about the development of this system is, as you know, water spray systems that are mounted on the cutter head of a tunneling machine. These are some of the most difficult to maintain and make operate, and they are operating at 100-200 lb/in^2. The reliability of those systems is notably low, so this is going to be a difficult development even though we are talking now of using moderate pressure compared to the ultra-high pressure that we have experimented with earlier.

The maximum rock hardness is somewhat a function of the diameter, particularly in economic tradeoff. The bigger the diameter, the slower the head has to turn, and in hard rock, you are getting a limited penetration per revolution. Therefore, the more difficult it is for a machine of, say, 40 feet or more in diameter to keep pace with the advanced rates that you could get with drill and blast. However, I would say that rock strengths are being bored economically today in moderate sizes of 15-20 feet in diameter, which are over 35 and up to 40,000 lb/in^2. There are very few tunnel jobs where you will find rock strengths higher than that. The rock conditions may be more limited factors, that is, the lack of joints and fractures to help you. The minimum practical size for hard rock is probably about 10 feet in diameter or for very hard rock, 12-13 feet in diameter. The maximum sizes depend on the condition of the rock. If it is very good rock, I do not see any problem with a 50-foot-diameter machine. The length of the tunnel depends on the economics. Usually today, existing machines, that is used machines, will be used on short tunnels. But that may depend on the cost of mobilization. If you have a tunnel that is only 2,000-3,000 feet long, especially in Europe, machines are often being taken off of a previous job and reused.

Immersed Road Tunnels in Open Sea Conditions

KAJ HAVNOE, Executive Vice President
Christiani & Nielsen A/S
Copenhagen, Denmark

STRAIT CROSSINGS

To discuss tunnels in open sea conditions is, for all practical purposes, synonymous with a discussion of strait crossings.

Figure 1 is a map of the world showing proposed major strait crossings, where immersed tunnels are a possible solution. Also included, however, is the Chesapeake Bay crossing executed between 1960 and 1964, as it is, so far, the only crossing which is executed in what may be termed open sea conditions.

Figure 1. The proposed major strait crossings.

To the right, we have the proposed 9.3-mile-long Tokyo Bay crossing. According to existing plans, it will be very similar to the Chesapeake crossing, as it will be a combination of bridges, artificial islands, and an immersed tunnel. The maximum water depth is 100 feet.

In the center of Figure 1 is the 1.9-mile-long Messina Strait crossing. Here, however, the maximum water depth is 500 feet. The likely solution is, therefore, a bored tunnel or a bridge. From time to time, the idea has been put forward of building a "floating tunnel," i.e., an immersed tunnel with positive buoyancy held down by wire cables attached to anchors in the bottom; in other words, more or less an inverted suspension bridge.

Tunneling and Underground Transport: Future Developments in Technology, Economics, and Policy
F.P. Davidson, Editor

The writer does not believe that, with the present technology, a traffic tunnel of such a design would be constructed, in view of the hazards to which the wire cables are subject--not least from corrosion.

A little to the west, we have the 15-mile-long Gibraltar Strait crossing. Here there is the standard discussion on the type of crossing: bridges, bored tunnels, classic immersed tunnels, and "floating tunnels." However, as the maximum water depth is 1200 feet, the crossing is likely to be built as a bridge or a bored tunnel.

The majority of the proposed strait crossings shown, have one factor in common; they have been talked about for over a century but nothing seems to be done to get them built.

There are, however, exceptions: (1) The English Channel, where both the British and the French Governments have recently issued their guidelines for a crossing, and (2) The Great Belt in Denmark, where the Danish Government is at present submitting its proposal to Parliament for a fixed crossing.

It would, therefore, now seem possible that these two crossings may be built within a foreseeable future.

For both crossings there is the traditional discussion of bored tunnels, immersed tunnels, and bridges, and a combination of bridges and immersed tunnels. At present, it seems likely that the Channel Tunnel will be a bored tunnel, while there is a fair chance that part of the Great Belt Crossing--under the Eastern Channel--will be an immersed tunnel.

ECONOMY

How to achieve an economical design is of course one of the designers' main concerns.

Figure 2 shows various features that will achieve economy. Starting from the shores, embankments should be used as far out as is economical and permissible from the point of view of hydraulics and environment. From there bridges with reasonably long spans, and piers secured against impact from ships should be constructed.

Then, artificial islands with ramps should follow. We have, in several cases, found that considerable economy can be obtained by constructing ramps with natural earth slopes--using a permanent ground water lowering system, as shown in Figure 3--instead of the concrete "bathtubs" shown above, which have to be kept stable against buoyancy forces (gravity ramps).

This requires, however, that an impermeable layer exists not too deep into the natural ground. Although the size of the structure with natural earth slopes is much larger than the "bathtub" structure, it consists mainly of cheap fill. As a matter of fact, in order to be able to build the "bathtub," a temporary structure very similar to the permanent structure for the drained ramp would be needed. Consequently, by adopting the drained ramp solution, a temporary structure is being upgraded into a permanent one.

Reverting to Figure 2, there is, then, the tunnel itself. Wherever existing hydraulic conditions permit, the tunnel should be placed above the deepest point in the strait. An optimum for the construction costs can be found when the maximum depth of the tunnel is reduced by the construction of an underwater embankment.

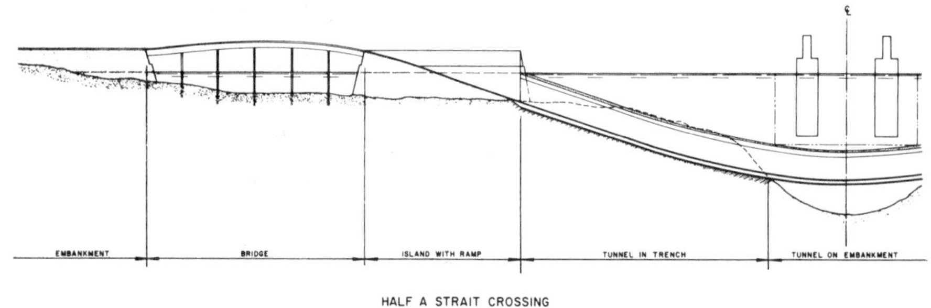

Figure 2. Typical elements in a strait crossing.

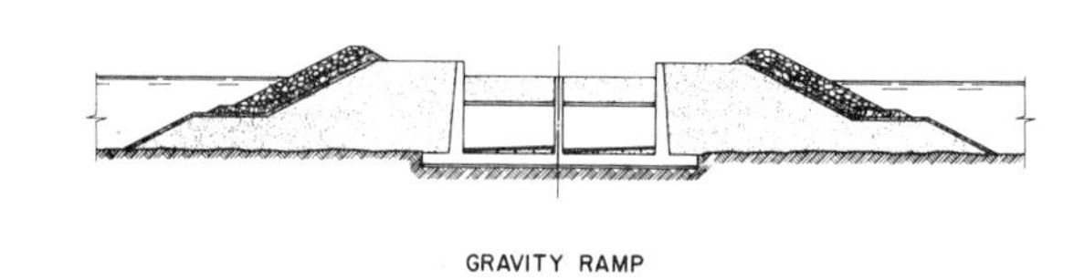

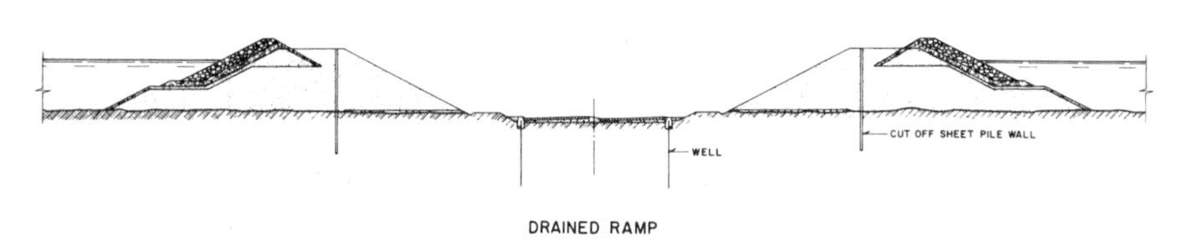

Figure 3. The cross sections of an artificial island with a ramp.

Strait crossings are generally for rural highways. As shown in Figure 4, it is typically a 4-lane tunnel with safety lanes and separate ducts for each direction.

The demand for cleaner exhaust from petrol cars--which originated in the United States--has spread all over the world. The PIARC recommendation on tunnel ventilation has halved the content of carbon dioxide and other toxic gases from the 1975 edition to the 1983 edition. This has resulted in the air requirement today being generally governed by the demand for visibility.

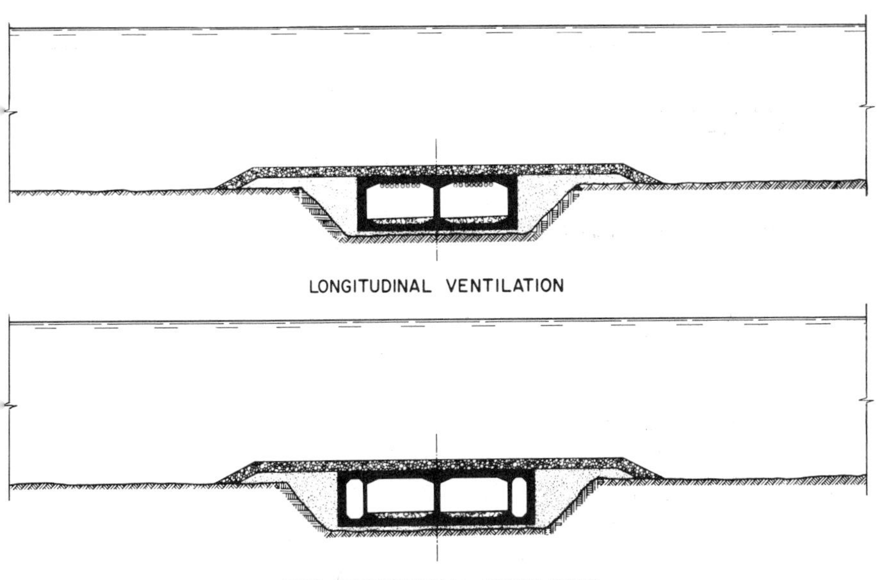

Figure 4. The cross sections of an immersed tunnel.

Depending on the gradients, a 2.5 to 2.8-mile-long tunnel can today be built with longitudinal ventilation without exceeding an air speed in the traffic duct of 37-39 ft/sec, which is considered the present limit. This type of ventilation--in the opinion of the writer--is also the safest in case of fire.

Compared with the more traditional transversal or semitransversal ventilation, there is a marked saving in cost for both civil works and the M&E installations. The two different cross sections shown on Figure 4 bear witness to this.

The length of 2.5-2.8 miles meets the demand of unhindered passage for shipping. If, however, local circumstances require longer tunnels, the writer advocates that tunnel designers look seriously into the trend for even cleaner exhaust--especially diesel exhaust.

By using computerized traffic control in the first years--until legislation picks up, it is possible to limit the maximum number of vehicles in the tunnel at any given time, and thus be able to choose the cheaper longitudinal ventilation. The writer is convinced that legislation will soon pick up, however.

An emission gas recirculation device designed to clean diesel exhaust is already in use for passenger cars. Debates on using this device on trucks also will be taken up by the European economic community this autumn. This will probably not yet result in any firm rules, but here again, the writer believes that the United States, and especially California, may well be ahead of Europe.

THE GREAT BELT CROSSING

Figure 5 shows the proposed combined rail and road tunnel under the 5.5-mile-wide Eastern Channel of the Great Belt in Denmark.

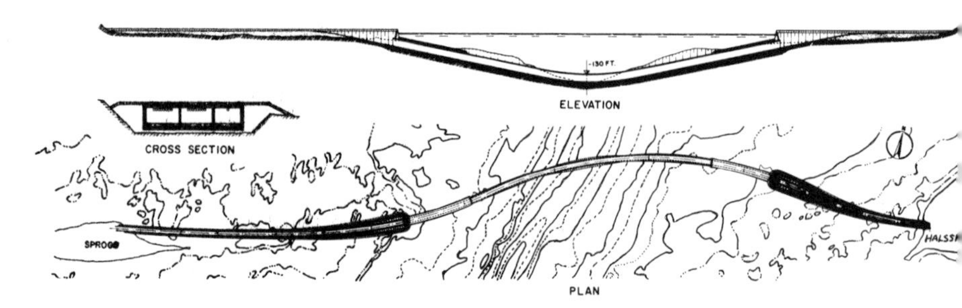

Figure 5. The proposed combined rail and road tunnel under The Eastern Channel of the Great Belt, Denmark.

This proposal includes most of the cost-saving features mentioned previously. There are two embankments, 1 mile and 2 miles long, respectively, ending in open ramps with natural earth slopes and a permanent ground water lowering. The embankments are carried out to a natural water depth of 33 feet.

There is, then, the 2.5-mile-long immersed tunnel, providing shipping with unhampered passage. By today's count, the number of ships passing each year amounts to 20,000, of which 97% are under 40,000 dwt. The maximum size so far has been a 396,000 dwt. tanker, albeit not fully loaded.

In the middle of the strait there is a 13-foot-high embankment. The tunnel has longitudinal ventilation.

This combined rail and road tunnel competes favorably, in regard to construction costs, with a bridge with a cable-stayed midspan of 2600 feet and a vertical clearance of 200 feet, and it has the added advantage that it will not be open to impact from ships, which is bound to occur in the case of a bridge. In fact, in the opinion of the writer, this should be sufficient reason for building a tunnel.

EXECUTION

In regard to execution, there are four main operations to be performed: dredging, foundation, sinking and joining, and back-filling.

Figure 6 shows how, in 1961, these operations for an immersed tunnel under the English Channel were visualized, i.e., under rather exposed conditions. All operations were to be carried out from walking jack-up platforms.

The writer thinks that this was a fairly risky proposal--it demands very calm weather to bring a tunnel element with, say 60.000 tons displacement so close to a jack-up platform--and probably many working days would be lost due to unfavourable sea conditions.

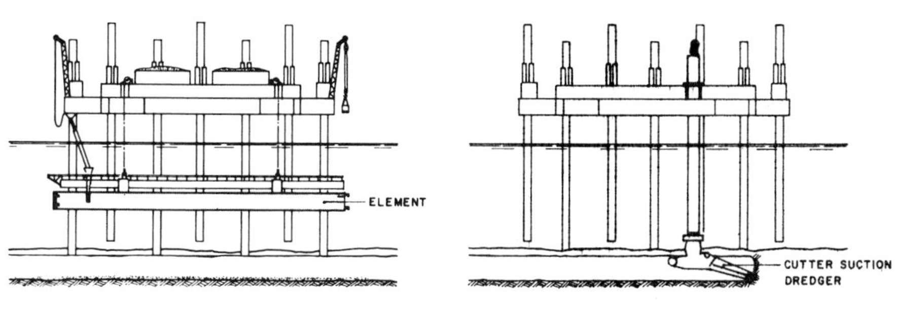

Figure 6. The 1961 proposal for execution of an immersed tunnel under the English Channel.

At that time my company, in order to reduce downtime because of sea conditions, prepared a sketch design where all operations, excavation, sinking, and a pumping of a sand foundation were carried out by means of a specially built underwater rig. The element to be sunk was hauled underwater, riding on top of the already executed tunnel.

Since then, however, there have been important developments in the field of semi-submersible oil-drilling rigs. Figure 7 shows how--benefiting from such experience--it could be possible to have the same operations carried out by a semisubmersible rig.

Figure 7. The 1985 proposal for execution of an immersed tunnel under the English Channel.

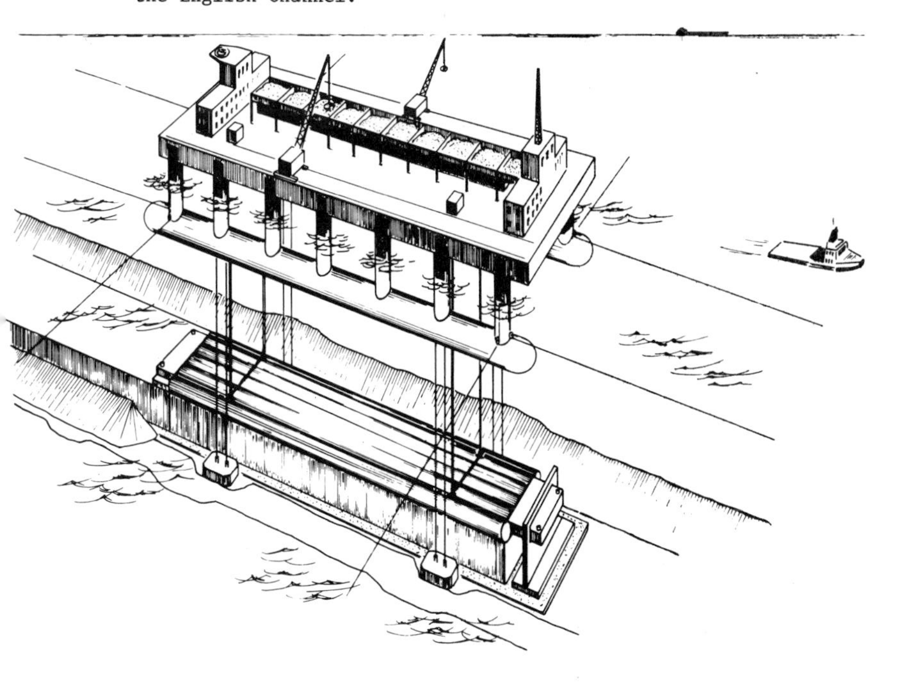

The rig is secured in the correct position by anchors. For the delicate screeding and sinking operations, the rig is vertically tethered. The vertical movements of the rig are eliminated by lowering four heavy anchor blocks to the bottom and tightening the vertical wire system. A frame for screeding of a gravel bed is towed in between the legs and lowered down, the bed is then screeded, and the frame removed.

The element, having a negative buoyancy of approximately 10%, is kept floating by buoyancy tanks, towed in between the legs, and suspended from the sinking tackles, which are furnished with heave compensators designed to reduce the suspension force oscillations.

The buoyancy tanks are filled with water, and the element is lowered and joined, using a rubber gasket for sealing between the elements. Finally, the element is brought in contact with the screed foundation. This is the situation shown in Figure 7.

Next, the buoyancy tanks are lifted out, deballasted, and towed away. Thereafter, the locking fill to half height of the element is placed by means of a chute from the rig, which can then be warped to the next position. The operation cycles should take two weeks.

What the writer is trying to convey is that he finds the combination of a fixed rig and large floating bodies very risky. It is advisable to use a floating element-laying rig, which should be so robust and well protected by fenders, that it can take a certain beating from unintended contact with the tunnel element.

The downtime, using a semisubmersible, will be less than with a jack-up platform, and certainly the risk for fatal loss of equipment will be greatly reduced.

ICBM Deep Basing Egress System Program

PHARO A. PHELPS
Manager of Defense Projects
Bechtel National, Inc.
San Francisco, CA 94119

The headquarters, Ballistic Missile Office in San Bernardino, California, has led U.S. Air Force studies of alternative means of basing Intercontinental Ballistic Missile (ICBM) weapon systems. These studies are aimed at insuring the ability of missiles to survive a nuclear attack, and guaranteeing their capability of retaliating. One alternative is known as deep basing. Under this concept, missiles would be stored deep underground from where an egress system would prepare tunnels and/or shafts to launch missiles. Such an underground base could probably qualify as the world's largest undertaking underground, therefore a brief review of one part of this concept is most fitting as a topic of interest to the American Society for Macro-Engineering.

In conjunction with the investigation of the deep basing alternative, the U.S. Air Force, through evaluation of competitive proposals, selected Bechtel National, Inc., of San Francisco, and the R.A. Hanson Company, of Spokane, Washington, in mid-1982, to assist independently with the development of an egress system for the missiles. These investigations of possible systems for providing egress from an underground base included both engineering studies and field demonstrations of the state-of-the-art capability to bore vertical shafts through rock and rubble. The Air Force has prepared a film* on the field demonstrations, and has graciously authorized its presentation at this conference. After the film, I will add a few words on the engineering study efforts.

I think it is safe to say that all concerned with these field demonstrations depicted in the film were satisfied with the results. Careful evaluation of the data gathered from our instrumentation system has led us to concepts for equipment modifications that we think will improve excavator performance, particularly in rubble.

Parallel with the field tests, we conducted engineering studies of potential egress systems. The concepts developed under the engineering study tasks were premised on use of state-of-the-art technology and equipment. Tunnel configuration, direction of egress (vertical, horizontal, inclined or a combination of the three), and the use of both manned and remote-controlled equipment were some of the issues studied in the engineering work, which addressed a variety of concepts that may have applicability to egress systems.

A multidisciplinary group of engineers screened and evaluated 24 conventional and unconventional excavation methods, the latter including the use of chemicals and lasers. A total of 9 systems handling the excavated material and 15 methods of ground support were studied, both alone and in various combinations. The initial screening study resulted

---

* Ed. note: A 13-minute U. S. Air Force film was shown at this point. The film documents the conduct and results of the demonstration tests which were conducted by the two prime contractors at the Department of Energy's Nevada Test Site. Requests concerning the film should be addressed to the Commander, U. S. Air Force Ballistic Missile Office, Norton Air Force Base, San Bernardino, California.

Published 1987 by Elsevier Science Publishing Co., Inc.
Tunneling and Underground Transport: Future Developments in Technology, Economics, and Policy
F.P. Davidson, Editor

in the selection of six potential methods of egress for further study, one of which involves the use of a full-face boring machine, a material handling system, and segmented steel liners--all of which would be stored underground with the missiles. Engineering studies screening for egress system candidates were conceptual in scope. They were a preliminary step toward more detailed and quantitative studies to be undertaken in future phases of the work.

It is obvious that a deep basing egress system should be able to excavate at a high advance rate. Our engineering studies and field test demonstrations both confirm what previous studies of major tunneling projects conclude--that the major causes of low advance rates lie with supporting systems, not the excavator. We think that an important lesson learned is that the systems engineering approach to the design of the total tunneling system shows promise of raising system availability as well as reliability and, therefore, advance rate.

For those of you who have an interest in a brief quantitative report of this work, which it has not been possible to offer here, I would like to refer you to the reports which were published last year in the Newsletter of the U.S. National Committee on Tunneling Technology, and to the paper published in the proceedings of the Rapid Excavation Technology Conference held in Chicago in June, 1983.

I would like to acknowledge the contributions of Bechtel's subcontractors and consultants who deserve full credit for their roles, which led to the successful results of our team. The Robbins Company manufactured the vertical boring machine, the Boeing Aerospace Company assembled and operated the data acquisition system, the Kenny Construction Company performed the horizontal boring, Dr. Levent Ozdemir of the Colorado School of Mines provided consulting on cutter studies, Carnegie Mellon Institute initiated a study on automating the egress systems, and Dr. Ed Cording of the University of Illinois, provided geotechnical consultation. I also wish to express my personal appreciation to the Bechtel National team for their leadership and dedication, and in particular to Mr. John Woolston, an MIT graduate by the way, who served as Deputy Program Manager and to Mr. Daniel G. Culver, who was the Deputy Program Manager for Field Construction. The success of both of the contractors was due in substantial part to the closely organized efforts of each contractor team, the Department of Energy's Nevada Test Site supporting personnel, and the U.S. Air Force team from the Ballistic Missile Office and the Air Force Regional Civil Engineer.

The Potential for Tunneling Based on Rock and Soil Melting

JOHN C. ROWLEY
Associate Director
Los Alamos National Laboratory, Los Alamos, NM

In 1960, several scientists at the Los Alamos National Laboratory, while working on quite another task, were impressed by the potential use of nuclear or electric power for melting rocks. None of us involved in the early discussions can remember just what tweaked our interest in the possibility of, if you will, transcontinental tunneling. We were concerned primarily with nuclear reactors and, of course, in this framework we started with the idea that long-range tunneling systems would logically have the major source of power within the tunneling machine itself. One could argue that point, but anyway, that is where we started. It was not until about ten years later that we actually found somebody interested enough in the idea to fund the basic research and engineering. The initial research which I shall describe was funded by the National Science Foundation.

The basis upon which we started was a little chart of melting points (Table 1). This chart showed refractory metals having melting points that are well above the melting points of most rocks. We thought the melting process was going to pose the most difficult problem. We discovered, however, that there are many feasible ways to melt rocks and soils. What really is difficult is handling the melt (Fig. 1). Since most of us have not delved deeply into tunneling technology, and because we were all systems-oriented people, we thought from the beginning about the melting process and the excavation as parts of a coherent system.

TABLE 1. Comparisons of Melting Points

Material	Melting Temperature °C	°F
Carbon	> 3500	> 6300
Tungsten	3380	6120
Boron Nitride	≃ 3000	≃ 5400
Molybdenum	2610	4730
Iridium	2443	4430
Kaolinite	1785	3245
Platinum	1769	3216
Quartz	1710	3140
Iron	1535	2795
Granite	1250	2280
Orthoclase	1200	2192
Balsalt	1150	2100
Albite	1110	2030
Gla	1100	2010
Aluminum	660	1200
Ice	0	32

Published 1987 by Elsevier Science Publishing Co., Inc.
Tunneling and Underground Transport: Future Developments in Technology, Economics, and Policy
F.P. Davidson, Editor

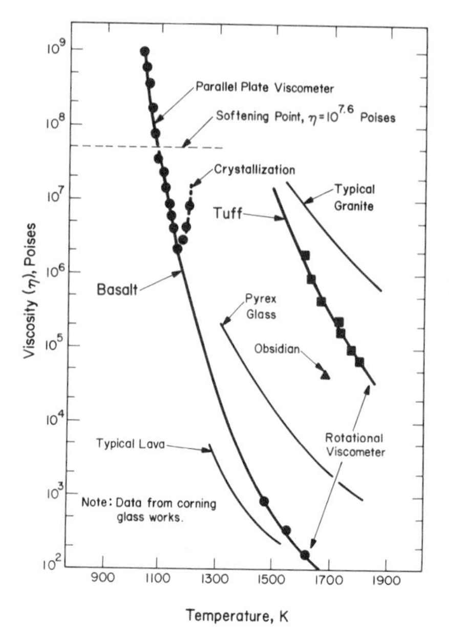

Figure 1. Viscosity of several rock melts.

Our main development tool was built to a relatively small scale. The reason we decided to take this approach was not only to reduce costs and time but also because we had a good group of analysts; as step succeeded step, they were very carefully performing the analytical calculations and perfecting computer programs for what we later called lithothermodynamics, that is, the fluid mechanics, the description of the melting process and the structural understanding needed to develop the hot, melting penetrators (Fig. 2). We tried to use conventional components wherever possible. The initial devices were about two inches or 50 millimeters in diameter.

Figure 2. Schematic diagrams of various modes and types of rock melting penetrations.

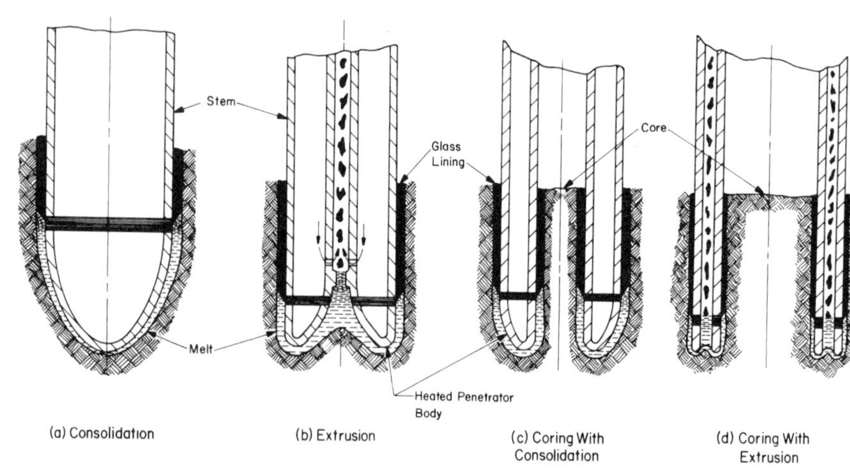

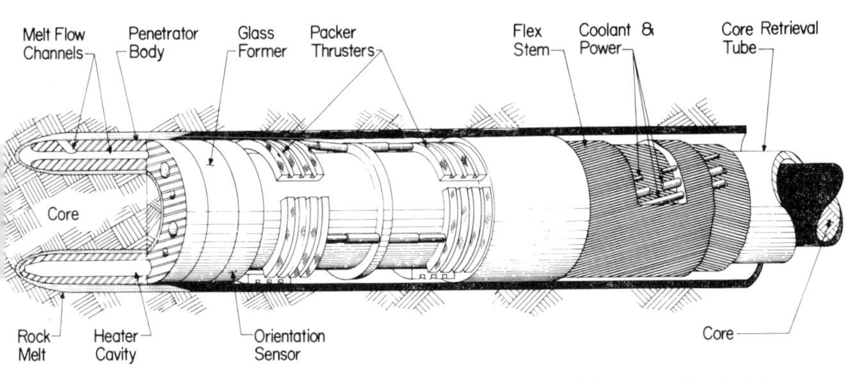

igure 3. Small diameter coring geoprospector, possible prototype to large tunneler concept.

Basic parameters of melting penetration were established by extensive aboratory experiments, field trials, and process modeling. Melting imitations, refractory metal fabrication, electric-powered heaters, melt andling, formation of glass linings, and debris handling were nvestigated. These results provided the scaling laws needed to prepare unneling machine prototype designs (Figs. 3-6).

There are special features of the melting process which should be mphasized. These include insensitivity to variable ground, borehole wall tabilization with glass linings (Fig. 7), long "bit" life, choice of ebris form (Figs. 8-10), and system automation.

igure 4. Alternate geoprospector design.

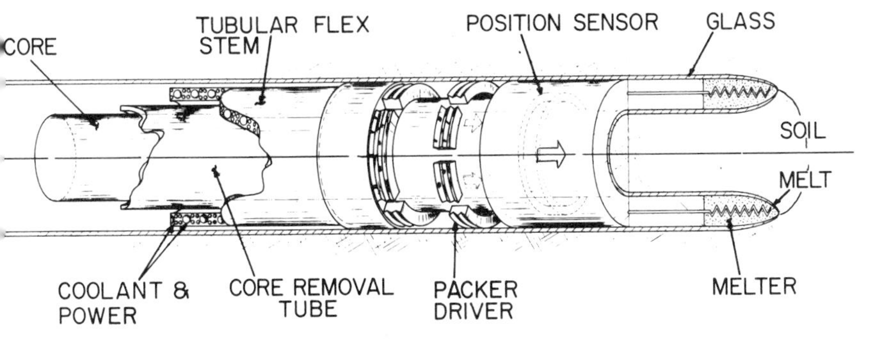

GEOPROSPECTOR

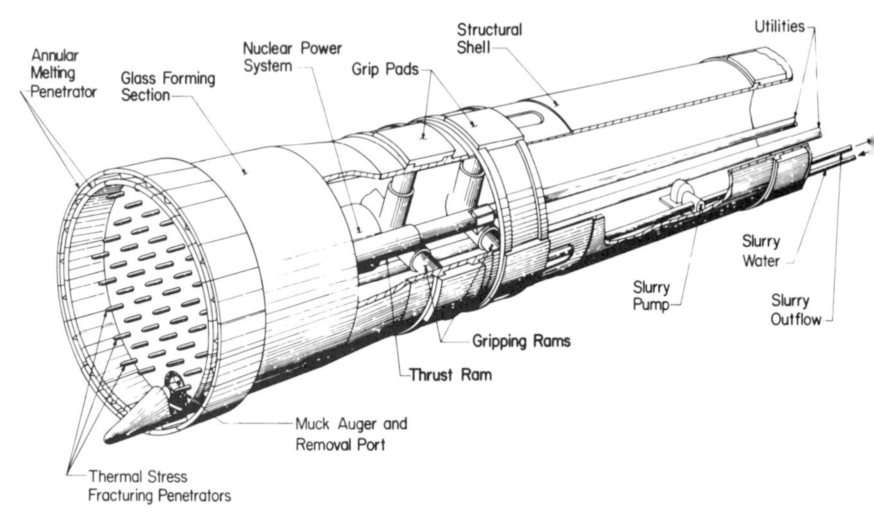

Figure 5. Hard rock, large diameter tunneler concept.

Figure 6. Soft ground, large diameter tunneler design concept.

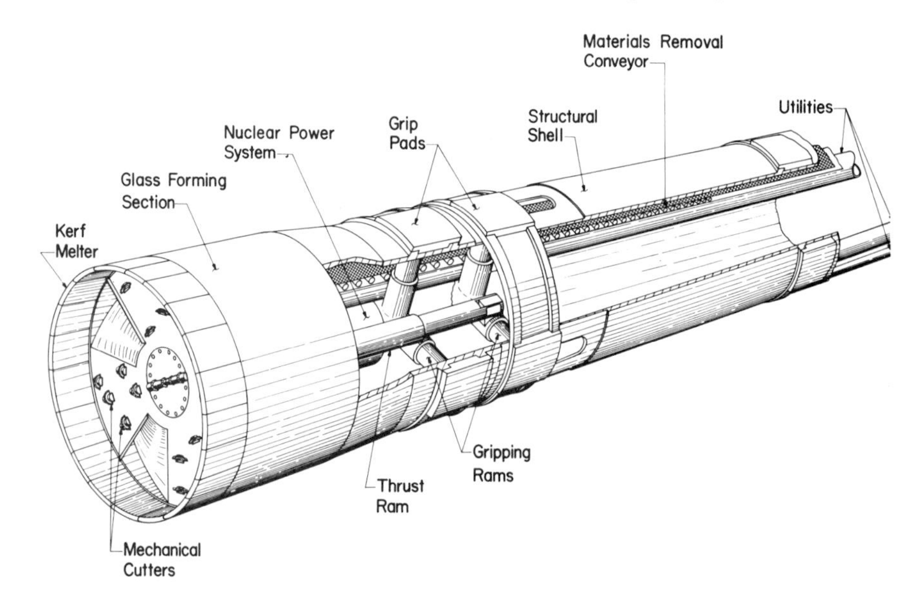

igure 7. Example of intact glass lining formed by consolidation in tuff.

igure 8. Typical glass rod debris.

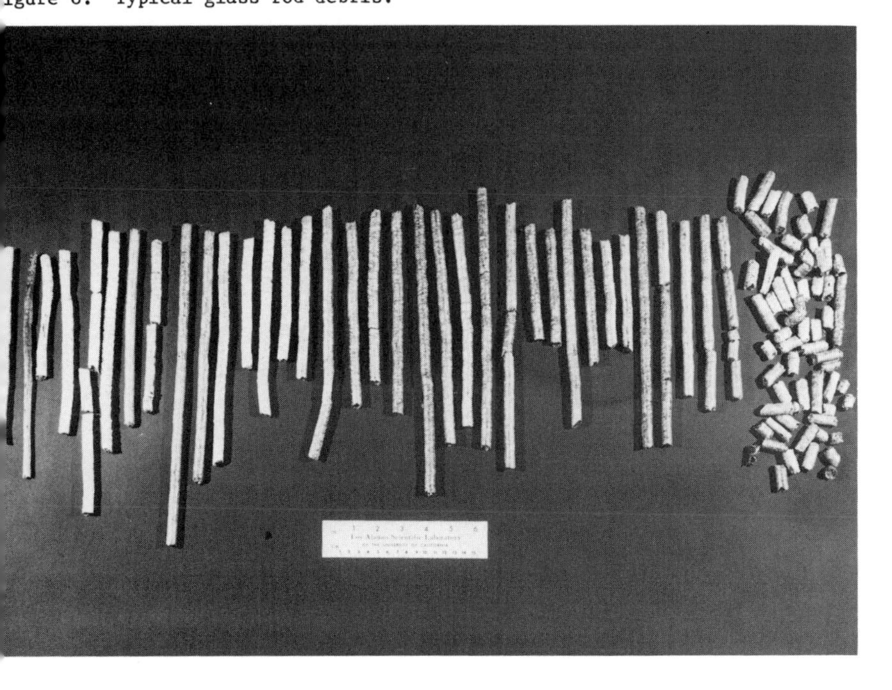

Figure 9. Glass pellets from basalt hole melting tests.

Figure 10. Mixture of glass pellets and glass-wool.

In May, 1975, a review of research progress was published under the itle, "The Subterrene Program" (Mini-Review 75-2), and it seems pertinent o quote from this document at some length:

> The Subterrene, invented and patented by scientists of the Los Alamos Scientific Laboratory (LASL), makes vertical or horizontal holes in rocks and soils by progressive melting, rather than by mechanical chipping or abrading. Rocks are mixtures of minerals and therefore have relatively low melting points--about 1450K (2150°F). Refractory metals, such as molybdenum and tungsten, have much higher melting points than this and are used as melting body materials... The rock-melt can be chilled to a glass and formed into a dense, strong, firmly attached hole lining. Therefore, permanently self-supporting holes can be made with a melting penetrator even with unconsolidated sediments... .
>
> ...A significant advantage of the technique is that the three major facets of excavation (rock fracturing, debris removal, and wall stabilization) are accomplished in a single, integrated operation... For porous rock or soft ground, the density consolidation Subterrene can be used to simplify the excavation process. In these materials, the glass lining, formed when the rock-melt cools, is significantly more dense and therefore occupies a smaller volume than did the original porous rock. By melting out to a diameter larger than that of the penetrator, the molten debris from the hole can be entirely consolidated in the glass lining, completely eliminating the necessity for removing debris.

Three reports are of special interest for potential designs of unneling machines based upon rock and soil melting. They are:

J.H. Altseimer, Systems and Cost Analysis for a Nuclear Subterrene Tunneling Machine: A Preliminary Study (LA-5354-MS)

J.W. Neudecker, Conceptual Design of a Coring Subterrene Geoprospector (LA-5517-MS)

R.J. Hanold, Large Subterrene Rock-Melting Tunnel Excavation Systems, a Preliminary Study (LA-5210-MS)

Indeed, a very comprehensive and detailed program of research and xperimentation was carried out by the Los Alamos team over a period of nany years. The list of publications at the end of this chapter will ndicate the scope of this effort. As an example of the direction of the nvestigative effort, I would like to cite a "Status Report" issued in ebruary, 1977, compiled by R.J. Hanold, and entitled "Rapid Excavation by ock Melting--LASL Subterrene Program, September 1973-June 1976." The bstract reads, in part:

> Research has been directed at establishing the technical and economic feasibility of excavation systems based upon the rock-melting (Subterrene) concept. A series of electrically powered, small-diameter prototype melting penetrators has been developed and tested. Research activities include optimizing penetrator configurations, designing high-performance heater

systems, and improving refractory metals technology. The properties of the glass linings that are automatically formed on the melted holes have been investigated for a variety of rocks and soils. Thermal and fluid mechanics analyses of the melt flows have been conducted with the objective of optimizing penetrator designs. Field tests and demonstrations of the prototype devices continue to be performed in a wide range of rock and soil types.

Clearly, a major new option has been offered to industry. But its application will require substantial investments in the full-scale machinery and auxiliary equipment needed for final demonstration, optimization, and utilization. Those of us connected with the research program will welcome serious inquiries. If, as a matter of public policy, or because of commercial opportunities within the private sector, there should be a decision to provide sustained support for a nationwide network of tunnels, then the alternative method developed at Los Alamos will provide a system, largely substantiated by research and practice, deserving of consideration. At a time when novel initiatives in technology are seen as essential components of a successful economic strategy, the Subterrene may yet have its "day in court."

INDEX OF TECHNICAL REPORTS AND PRESENTATIONS

Copies of reports listed below can be obtained from:

National Technical Information Service (NTIS)
U.S. Department of Commerce
5285 Port Royal Road
Springfield, VA 22151

The completed reports are identified by their LA-MS number by NTIS.

Discussions of the technical reports can be directed to:
John C. Rowley
University of California
Los Alamos National Laboratory
P.O. Box 163, MS D462
Geology and Geochemistry Group
Los Alamos, NM 87545
Telephone: (505) 667-1378
(FTS) 843-1378

A. COMPLETED LASL TECHNICAL REPORTS

LASL Report No.	TITLE	AUTHOR(S)
LA-5354-MS	Systems and Cost Analysis for a Nuclear Subterrene Tunneling Machine - A Preliminary Study (September 1973)	J.H. Altseimer
LA-5422-MS	A Versatile Rock-Melting System for the Formation of Small-Diameter Horizontal Glass-Lined Holes (October 1973)	D.L. Sims

LASL .eport No.	TITLE	AUTHOR(S)
A-5423-MS	Carbon Receptor Reactions in Subterrene Penetrators (October 1973)	W.A. Stark,Jr. M.C. Krupka
A-5435-MS	Rapid Excavation by Rock Melting--LASL Subterrene Program--December 31, 1972 to September 1, 1973 (November 1973)	R.J. Hanold
A-5211-MS	Subterrene Electrical Heater Design and Morphology (February 1974)	P.E. Armstrong
A-5502-MS	Heat Transfer and Thermal Treatment Processes in Subterrene-Produced Glass Hole Linings (February 1974)	A.C. Stanton
A-5517-MS	Conceptual Design of a Coring Subterrene Geoprospector (February 1974)	J.W. Neudecker
A-5540-MS	Selected Physiochemical Properties of Basaltic Rocks, Liquids and Glasses (March 1974)	M.C. Krupka
A-5573-MS	Development of Mobile Rock-Melting Subterrene Field Unit for Universal Extruding Penetrators (April 1974)	J.E. Griggs
A-5608-MS	Numerical Solution of Melt Flow and Thermal Energy Transfer for the Lithothermodynamics of a Rock-Melting Penetrator (May 1974)	R.D. McFarland
A-5613-MS	The AYER Heat Conduction Computer Program (May 1974)	R.G. Lawton
A-5621-MS	PLACID: A General Finite-Element Computer Program for Stress Analysis of Plane and Axisymmetric Solids (May 1974)	R.G. Lawton
A-5689-MS	Geothermal Well Technology and Potential Applications of Subterrene Devices--A Status Review (August 1974)	J.H. Altseimer
A-5826-MS	Characterization of Rock Melts and Glasses Formed by Earth-Melting Subterrenes (January 1975)	L.B. Lundberg
A-5857-MS	Chemical Corrosion of Molybdenum and Tungsten in Liquid Basalt, Tuff, and Granite with Application to Subterrene Penetrators (February 1975)	W.A. Stark,Jr.
A-5838-MS	Petrography and Chemistry of Minerals and Glass in Rocks Partially Fused by Rock-Melting Drills (September 1975)	S.N. Ehrenberg P. Perkins M.C. Krupka
A-6038-MS	Unique Refractory Techniques for Fabricating Subterrene Penetrators (September 1975)	W.C. Turner

LASL Report No.	TITLE	AUTHOR(S)
LA-6135-MS	Rock Property Measurements Pertinent to the Construction of Drainage Systems at Archeological Sites in Arizona by Subterrene Penetrators (November 1975)	G.M. Pharr
LA-6265-MS	Development of Coring, Consolidating, Subterrene Penetrators (March 1976)	H.D. Murphy J.W. Neudecker G.E. Cort W.C. Turner R.D. McFarland J.E. Griggs
LA-6555-MS	Technical and Cost Analysis of Rock Melting Systems for Producing Geothermal Wells (November 1976)	J.H. Altseimer

---

B. TECHNICAL PRESENTATIONS AND JOURNAL ARTICLES

Conference on Research in Tunneling and Excavation Technology (Abstract and Presentation), NSF, Wayzata, MN, September 14-15, 1973.
J.C. Rowley, Rapid Excavation by Rock Melting

15th Symposium on Rock Mechanics (Presentation and Paper), U.S. National Committee on Rock Mechanics, Custer, SD, September 17-19, 1973.
R.J. Hanold, The Subterrene Concept and its Role in Future Excavation Technology

26th Pacific Coast Regional Meeting (Abstract and Presentation), American Ceramic Society, San Francisco, CA, October 31-November 2, 1973
M.C. Krupka, Refractory Material and Glass Technology Problems Associated with the Development of a Rock Melting Drill

Tunnels & Tunnelling Magazine (Invited Article), British Tunnelling Society, January-February, 1974.
J.H.Altseimer, Subterrene Rock Melting Devices

University of Wyoming (Invited Presentation), Laramie, WY, April 25, 1974.
J.C. Rowley, Rock Melting and Geothermal Energy

Geotechnical Eng. Group and Association of Eng. Geologists Joint Meeting (Invited Presentation), Los Angeles, CA, May 30, 1974
R.J. Hanold, The LASL Subterrene Concept

Rapid Excavation & Tunneling Conference (Presentation and Paper), American Institute of Mining Engineering, San Francisco, CA, June 24-27, 1974
J.C. Rowley, R.J. Hanold, C.A. Bankston, and J.W. Neudecker, Rock Melting Subterrenes: Their Role in Future Excavation Technology

Petroleum Engineer (Journal Article), July, 1974
D.L. Sims, Melting Glass-Lined Holes: New Drilling Technology

ournal of Vacuum Science and Technology (Article), American Vacuum
ciety, Vol. 11, No. 4, July-August, 1974
W.A. Stark, Jr., et al, Application of Thick Film and Bulk
Coating Technology to the Subterrene Program

rd International Congress ISRM (Presentation and Paper), International
ociety for Rock Mechanics, Denver, CO, September 2-7, 1974
J.C. Rowley, Rock Melting Applied to Excavation and Tunneling

TO Committee on Challenges of Modern Society (Presentation), NATO, Los
lamos, NM, September 18, 1974
C.A. Bankston, The Los Alamos Scientific Laboratory Subterrene
Project and Its Applications to Geothermal Energy

onference on Research for Development of Geothermal Resources
Presentation and Paper), NSF, CPL, CIT, Pasadena, CA, September 23-25,
974.
J.C. Rowley, Rock-Melting Technology and Geothermal Drilling

ASA-Houston Technology Transfer Conference (Presentation and Paper),
ouston, TX, September 24-25, 1974
R.J. Hanold, C.A. Bankston, J.C. Rowley, and W.W. Long, The
Initiatives of the Los Alamos Scientific Laboratory in the
Transfer of a New Excavation Technology

arth & Planetary Sciences Group--Johnson Space Flight Center
Presentation), NASA, Houston, TX, September 26, 1974
R.J. Hanold, The Los Alamos Subterrene Program and its Role in
Geothermal Energy Development

7th Annual Meeting, Association of Engineering Geologists (Abstract and
resentation), AEG, Denver, CO, October 18, 1974
C.A. Bankston and J.H. Altseimer, The Rock Melting Subterrene
and its Potential Role in Geothermal Energy

IAA/SAE 10th Propulsion Conference (Presentation and Paper), AIAA/SAE,
an Diego, CA, October 21-23, 1974
J.H. Altseimer, J.D. Balcomb, W.E. Keller, and W.A. Ranken,
Nuclear Propulsion Technology Transfer to Energy Systems

974 ASME Winter Annual Meeting (Presentation and Paper), ASME, New York,
Y, November 18-22, 1974
R.D. McFarland and R.J. Hanold, Viscous Melt Flow and Thermal
Energy Transfer for a Rock-Melting Penetrator

unnels & Tunnelling Magazine (Article), British Tunnelling Society,
anuary-February 1975
R.E. Williams, Soil Melting--A Practical Trial

975 ASME Winter Annual Meeting (Presentation and Paper), ASME, Houston,
X, November 30-December 4, 1975
H.N. Fisher, Thermal Analysis of Some Subterrene Penetrators

niversity of Colorado (Invited Presentation), Boulder, CO, December 2,
975
J.H. Altseimer, The Subterrene Program and Geothermal Energy

C. REPORTS RELATED TO SUBTERRENE TECHNOLOGY PUBLISHED BY OTHER ORGANIZATIONS

1. Black, D.L., Basic Understanding of Earth Tunneling by Melting. Prepared for U.S. Department of Transportation by Westinghouse Astronuclear Laboratory, July 1974.

2. Bledsoe, J.D., J.E. Hill, and R.F. Coon, Cost Comparison Between Subterrene and Current Tunneling Methods. Prepared for National Science Foundation by A.A. Mathews, Inc., May 1975.

3. Black. D.L., A Study of Borehole Plugging in Bedded Salt Domes by Earth Melting Technology. Westinghouse Astronuclear Laboratory, June 1975.

4. Nielsen, R.R., A. Abou-Sayed, and A.H. Jones, Characterization of Rock-Glass Formed by the LASL Subterrene in Bandelier Tuff. Terra Tek, November 1975.

5. Muan, A., Silicate-Metal Reactions with a Bearing on the Performance of Subterrene Penetrators. The Pennsylvania State University, August 1976.

6. St. John, C.M., Stresses and Displacements Around Deep Holes in Hot Rocks. University of Minnesota, September 1976.

se of Submerged Caissons in the Construction of ndersea Tunnels

ILLIAM L. BOHANNAN
and
. VINCENT HARRINGTON
? Taurio Corporation
roton, Connecticut 06340

This presentation deals with providing the advantages of intermediate
ntry points to speed tunnel construction when these points are
nderwater. Intermediate entry points to tunnel construction sites are
ertainly not new; however, providing them along the underwater portion of
unnels is not very common.

The advantages of providing intermediate shafts for tunnel construction
ctivities are well known. Briefly, these shafts provide for faster
xcavation since more faces can be worked and they provide the contractor
ith more flexibility and less congestion at the various construction
ites. These factors tend to shorten construction time, which in turn
esults in less costly projects for several reasons; e.g., interest on
orrowed funds is minimized and revenue flow from the project can begin
ooner. The increased flexibility also reduces schedule risk, hence tends
o minimize chances for overruns.

The decision to use these underwater shafts and where they should be
ocated involves many variables. In addition to those variables which
ust be considered for a shaft sited on land, surface ship traffic
atterns, depth of water, currents, and bottom geology must be
onsidered. The planned method of handling the material removed from the
xcavation is also an important consideration in choosing the site. Even
he tunnel profile may enter into the decision.

To illustrate some of the unique factors that must be considered when
hoosing a site, I will use the English Channel Tunnel as an example.

This tunnel is projected to go beneath the English Channel between
over, England and Calais, France. The tunnel complex will consist of two
ail tunnels of 7.1-meter inside diameter flanking a 4.5-meter-diameter
ervice tunnel at 15-meter centers. This complex will be cross connected
y passages (adits) at 375-meter intervals.

The approximate route the tunnel will take is shown as an overlay on a
autical chart in Figure 1. The length of the tunnel will be about 50
ilometers. About 36 kilometers will be under the channel waters.
onsidering the surface ship traffic patterns, it would be best to locate
ny midchannel shafts with their associated surface activity outside the
ain shipping channels. Good locations from this point of view would be
n the dividing line between the inshore traffic zones and the main
hipping channels. These dividing lines are shown as shaded areas on
igure 1. If located at these points it will be possible to implant and
ervice the tunnel construction shafts without significantly interfering
ith either inshore or main channel ship traffic. Additionally, sites on
he dividing lines will be more likely to make it acceptable to use a
jack up tower" at the site to support the implanting of a shaft and the
ubsequent flow of construction materials through the shaft.

With the plentiful supply of water readily available for undersea
unnels, using slurry as a means of transporting the muck would probably

nneling and Underground Transport: Future Developments in Technology, Economics, and Policy
P. Davidson, Editor

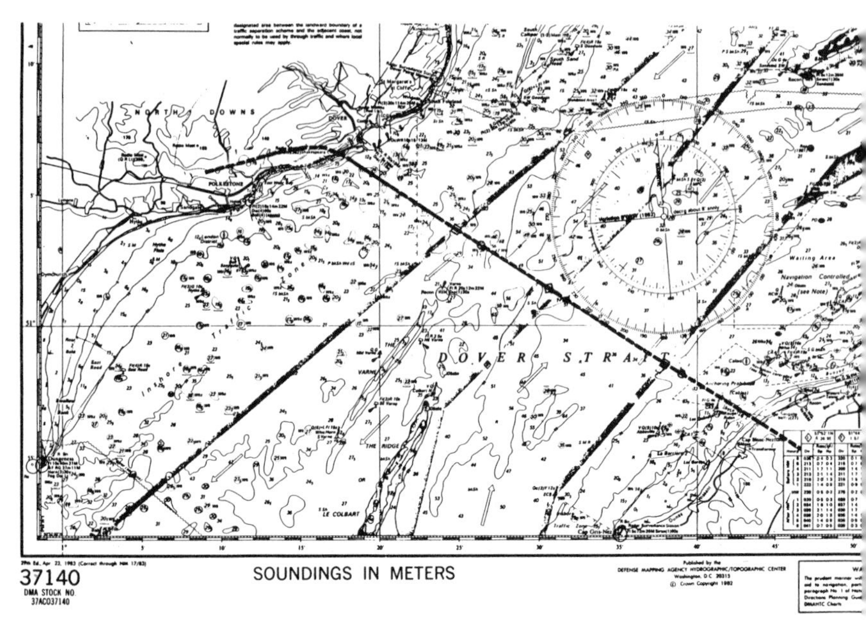

Figure 1

be a consideration. A slurry pipeline handling the muck within the tunnel bores would have the additional advantages of making the bores less congested during the construction while providing the means to move the material away from the faces on a continuous basis, therefore not restricting the speed of advance of the TBMs because of the rate we can muck. Long tunnels, such as the English Channel Tunnel, however, would require the handling of a considerable amount of pipe through the shaft. To preclude this, another concept that involves digging a flume in the invert of the tunnel bore to transport the slurried muck within the confines of the tunnel bore has been devised. These flumes would carry the muck slurry to the low points via gravity. Calculations show that indeed this scheme may be possible in the channel tunnel and would result in substantial savings of power and would also involve much less equipment to handle materials within the tunnel bores since gravity will provide the power and extending a slurry pipeline would not be necessary. If the low points of the profile could be made to coincide with the sites where we put the shafts, the slurry could be pumped to the surface and either loaded in barges or pumped from the top of the shaft to the disposal site in an underwater pipeline. In our channel tunnel example, the presently planned low points, W-1 and W-2'are positioned as shown by the small circles in Figure 1. By adjusting these points slightly it would be possible to place these low points such that they coincide with the boundaries of the established shipping lanes to achieve the advantages pointed out earlier. The relocated points are shown as small squares in Figure 1. In this case, relocation would be possible and still accommodate a tunnel profile which stays in the chalk marl layer, thus satisfying the intent of the existing profile.

Having reviewed some of the considerations which would be involved in site selection for the in-water shafts, let us now examine the concepts which are required to support the excavation of a shaft through the seabed and some ways that materials would be supplied to the tunnel construction

fforts through the water interface and shaft. Figure 2 shows two such
oncepts. The one on the right would use a cofferdam arrangement which
xtends above the surface and would be supported by a jack up tower rig
hroughout the construction period. Should water depths or other factors
reclude the use of the jack up tower, the concept on the left might be
sed by providing submersible barges for resupply. In both cases, a
ubmerged caisson of some design would be required to maintain the
atertight integrity and safety of the tunnel complex below.

This being the case, a caisson design and method of implanting it into
he sea floor will be necessary. Again using the English Channel Tunnel
s an example, the following sequence and design considerations would
pply. The caisson design process would first consider the amount of room
hat would need to be provided to support the tunneling operations and the
peration of the caisson systems themselves. The caisson size would be
ependent on the expected materials, staging areas, equipments, resupply
ethod, and types of materials that would be required to provide a good
ogistics flow during the construction. For our example, a caisson of the
ize and shape shown in Figure 3 would probably satisfy most criteria.
he "doughnut" with the 70-feet outside diameter, 36-feet-diameter hole,
nd 15-feet high would provide the needed volume for equipment and
aterial stowage and staging, slurry pumps, emergency, and electrical
upplies as well as caisson systems and a control station. The 50-feet-
iameter portion, which comprises the lower section of the caisson, will
onsist principally of variable ballast tanks which can be flooded or
umped. This will provide the necessary means of ballasting the caisson
or stability while floating it to the site and for buoyancy control while
mplanting the caisson.

igure 2

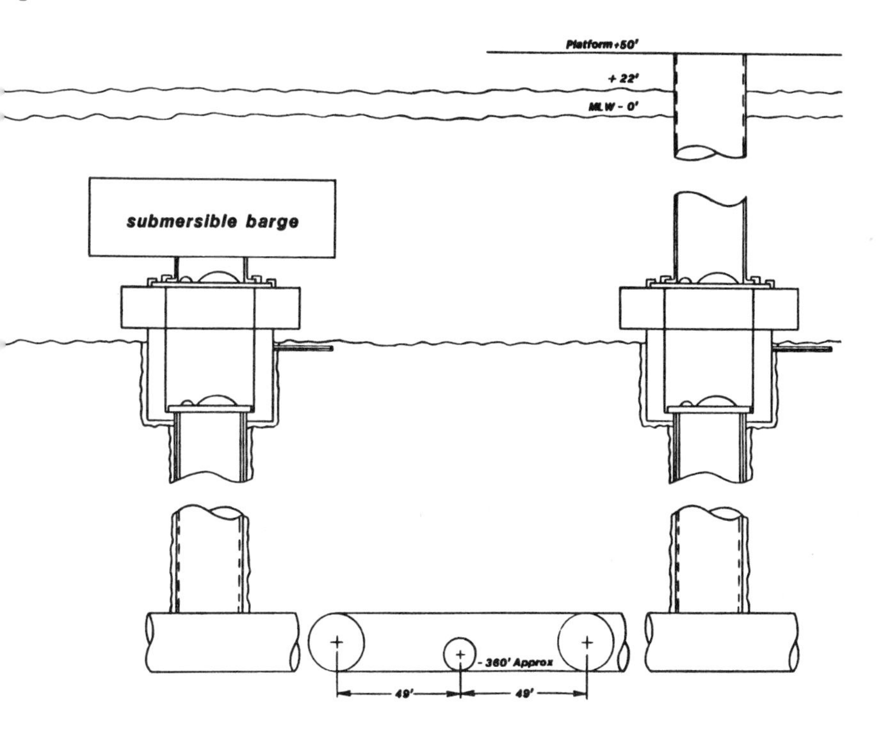

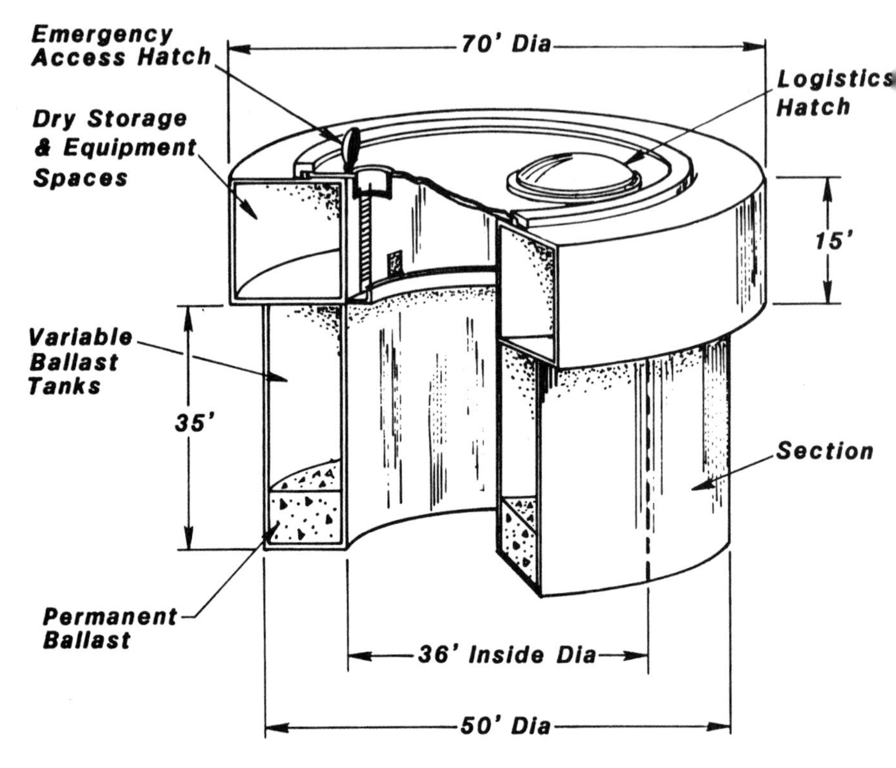

Figure 3

Once implanted these tanks can be flushed and used for storage of fresh water for concrete if required. The caisson will be built at a facility on shore, the initial load of construction materials and equipment put aboard in the dry storage areas. The caisson will then be ballasted for the trip to the site where it will be implanted. At the site a 54-feet-diameter hole will be dredged into the sea floor to a depth of 30 feet. The caisson will be towed to the site, ballasted down, and lowered into the dredged holes shown in Figure 4. Once it is leveled it will be grouted into place. Next, a drilling device will be lowered and mounted to the top of the caisson as shown in Figure 5. The drill will be powered by an oil well drill on the platform or by a drilling ship. A 28-foot-diameter shaft will be bored down to tunnel level, then the drill rig will be removed. Next, caisson section number 2, shown on Figure 6, will be barged to the site, lowered into the bored shaft using cranes from the jack up tower. Section 2 seats on a flange at the bottom of section 1. Once section 2 is positioned, it will be grouted to the sides of the shaft. Notice that section 2 has hatches on the top shown by the little humps. The cover of section 1 containing watertight hatches will next be lowered and secured into place. Once this cover is in place, the annulus of the caisson can be pumped dry. Next, connections to any underwater pipes, conduits going ashore, etc., would be made up, and if the tower is to be used, the cofferdam section will be installed and dewatered. The entire complex is shown in Figure 7. The caisson could then be manned and all systems tested.

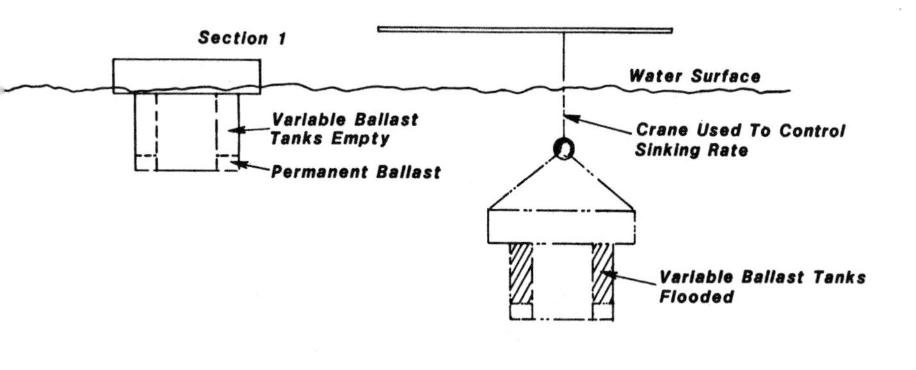

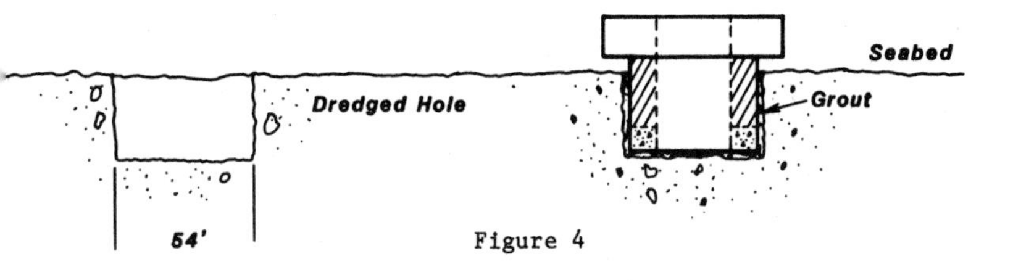

Figure 4

Figure 5

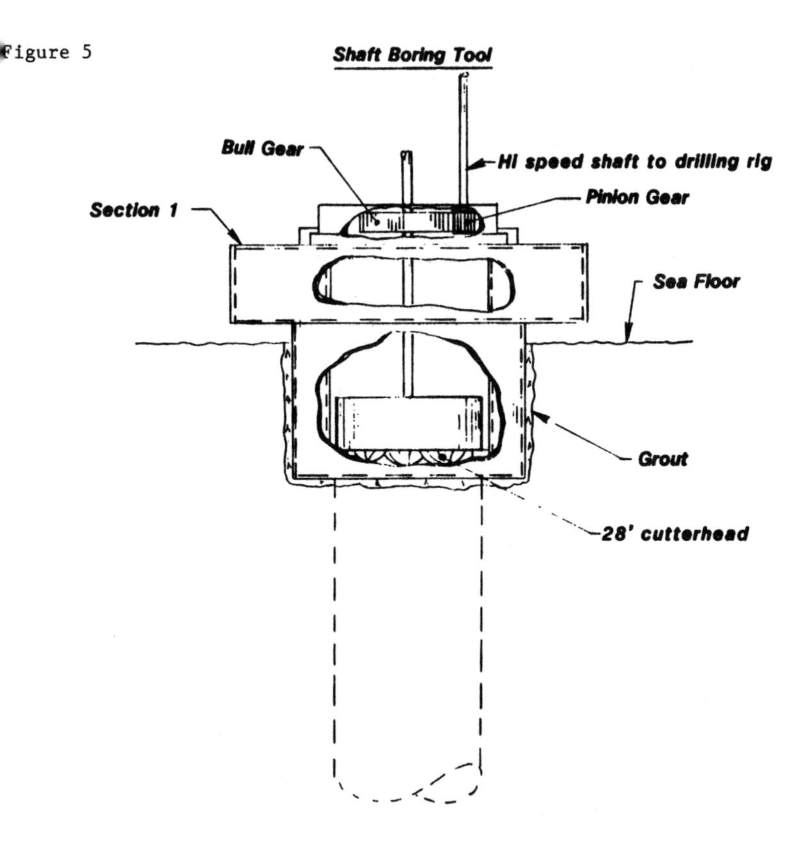

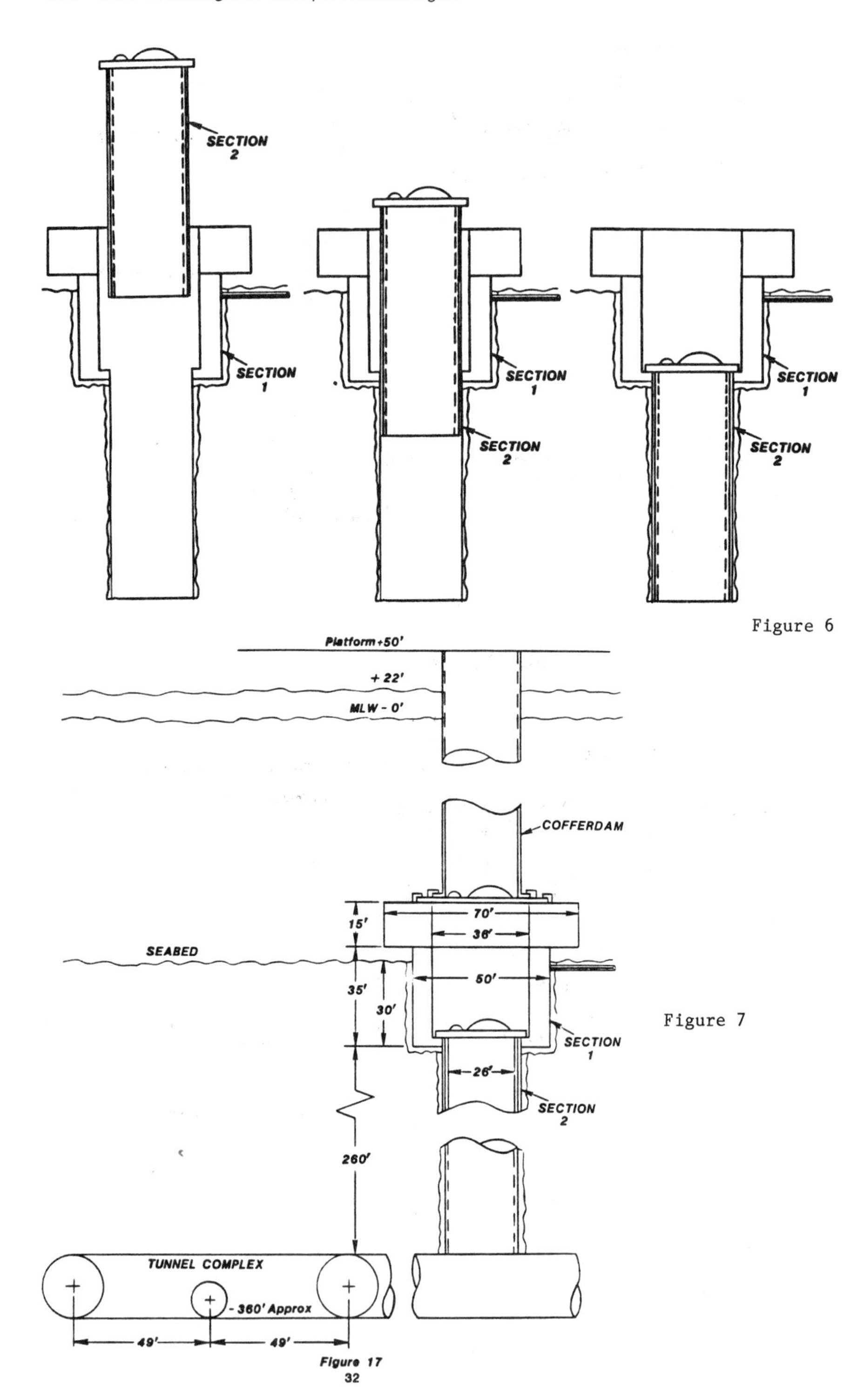

Figure 6

Figure 7

Figure 17

Since the caisson is intended to provide all manners of logistics and life support to the subterranean works, it must do so at a rate commensurate with the activities and progress. Hence the caisson design must reflect some of the following requirements: It must provide access from the work site to the sea surface and above for personnel, equipment, and materials; it must also provide for continuous power to machinery, safety, convenience, and hotel requirements; it certainly must accommodate at a near-continuous supply of materials for lining the exposed tunnel surfaces; and last but not least, it must accomodate a continuous removal of excavated material.

In the process of making these initial estimates for the size of the caisson and associated systems, some parametric studies have been done. Calculations relative to using a flume, excavated into the bottom of the tunnel for in-bore mucking, have been made.

The flume that would be required is not very big. The muck will be transported in the flume to the bottom of the shafts as a slurry. At this point we will pump it in pipes to the surface, either load it into barges or pump it via an underwater slurry pipeline to the disposal sites ashore. A flume of 6-square foot cross section will support mucking rates which equate to excavating the three faces at an advance rate of 12 feet per hour. Figure 8 shows the rate slurry would be produced plotted against the number of faces being worked and slurry density. Slurry production rates are shown in cubic feet per second since these units are very handy in computing flume and piping sizes. The group of bars on the left equate to working three faces, the two 8-meter faces and one 5-meter-diameter face. This would approximate the muck to be carried in the flume when all three faces were working. The group of bars on the right equate the muck production from six faces of the caisson. This would equate to the amount of muck which would have to be pumped out through the slurry risers from the bottom of the caisson. Figure 9 shows the capacity of a flume having the shape of the small diagram in the upper left-hand corner. The capacity is plotted against the depth of the slurry in the trench. Each curve represents the slope of the flume as shown by the key below. These slopes cover the ranges of slopes in the current profile plan of the English Channel Tunnel. Note that these curves are calculated assuming the flume is lined with shotcrete. Figure 10 shows the same set of curves, but this time we assume the flume is lined with a smooth surface such as plastic. This would be extruded immediately behind the tunnel boring machine. Such a concept would provide a very promising way in which we could begin mucking as a slurry right behind a tunnel boring machine. Actually, the flumes may make it possible to muck the tunnel at a faster rate than we could by using more conventional means from the shore sites.

Of course, getting material into the caisson must also be a major consideration and Figure 11 shows examples assuming a platform is installed and serviced to support the resupply. Figure 12 shows some factors to be considered if tower support is not available. The job is harder, but definitely still surmountable. Services for the caisson, such as compressed air, electricity, fresh water, and communication lines will all have to be supplied to the caisson. Electrical power will be required in substantial amounts. Figure 13 is an additional example of parametric studies that we have performed to estimate this power need. It shows the pump power required to pump the slurry from the tunnel level to the top of the caisson with all six faces of the tunnel being worked simultaneously. The impact of the slurry density on the pumping power is very evident. We would want to work probably between 1.2 and 1.3 density. Also included are the various pipe sizes and the numbers or risers of various pipe sizes

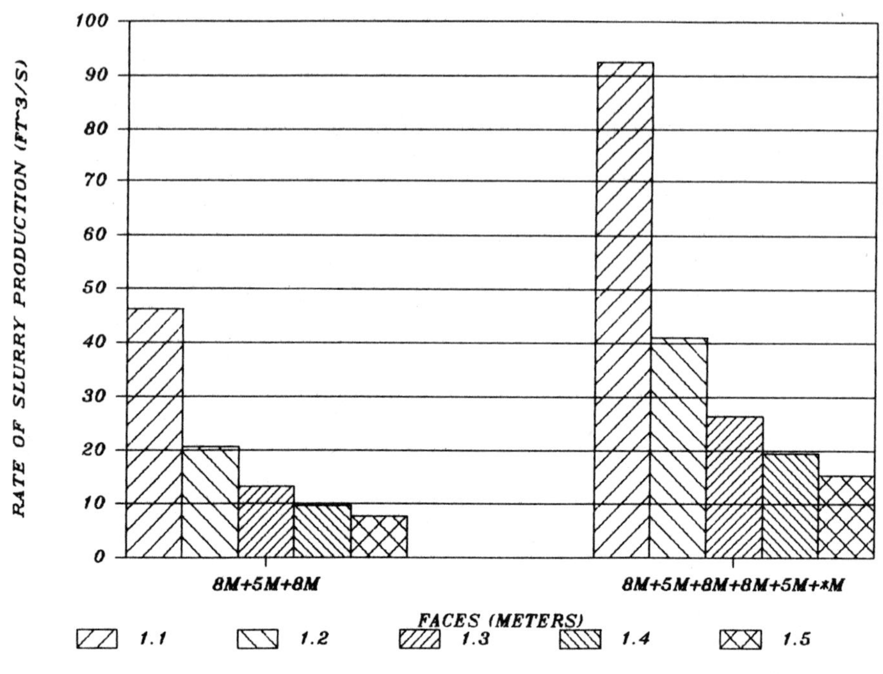

Figure 8

Figure 9

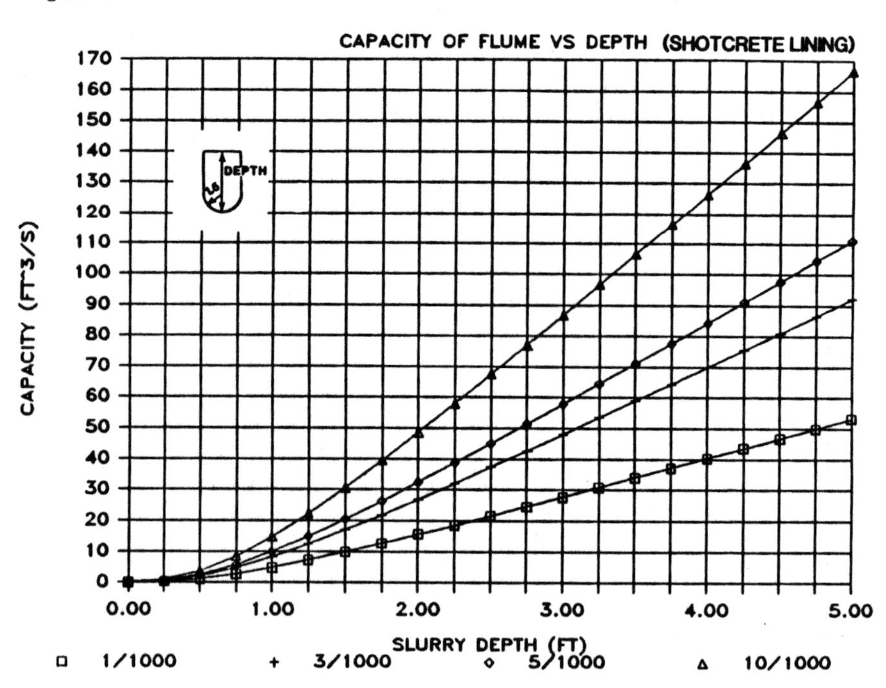

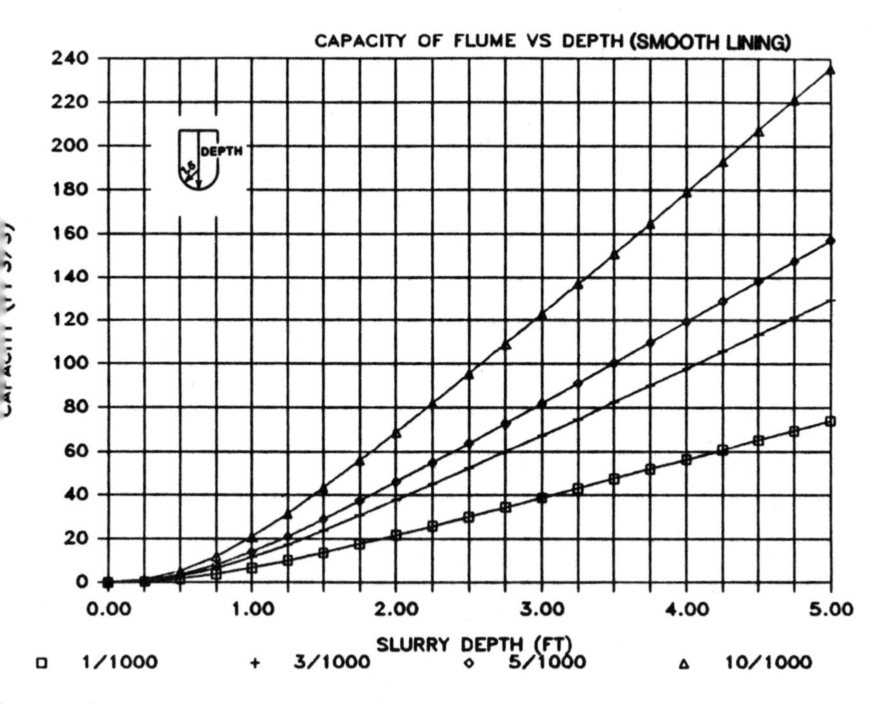

Figure 10

Figure 11

Getting Material into the Caisson

- ***Easiest solution would be to use a jack up platform to support each caisson site.***
  - ***This would permit using a cofferdam extending from the top of the caisson to the surface.***
  - ***Off duty crews would use the platform facilities.***
  - ***Vertical lift assets of the platform could then be used for the lions share of the vertical traffic.***
  - ***Platform would be used as a staging area and hence could provide the requisite flexibility in scheduling the delivery of materials.***
  - ***Dry fines, liquids would be handled through piping.***
  - ***Pipe casings could be lowered in strings through blowout protection devices.***
  - ***Bulk materials such as wire matting, rock bolts, etc can be lowered in freight transfer boxes.***

- ***If a tower rig is not practical, resupply can be accomplished using a submersible barge which would mate to the top of the caisson.***

  - ***Barge would be used to ferry the material from shore site to caisson.***

  - ***Hotel accommodations for work crews would be designed into the caisson.***

  - ***Staging of materials would be accomplished in the caisson or in subterranean spaces at tunnel level.***

  - ***Vertical lift capability would be intergral to the submersible barge and more lift capability would be designed into the caisson.***

Figure 12

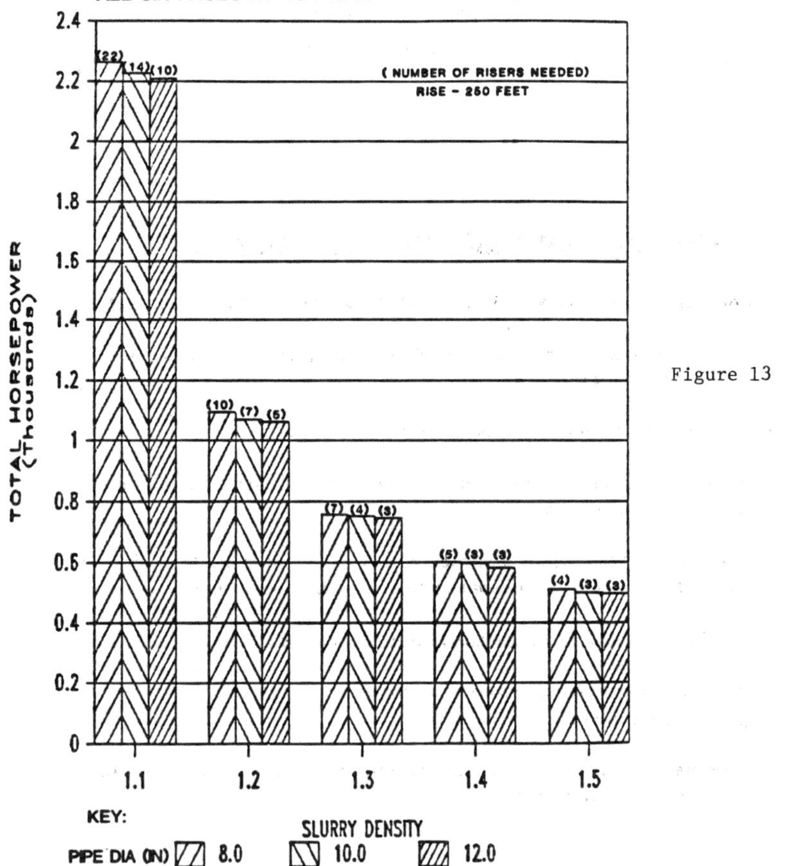

Figure 13

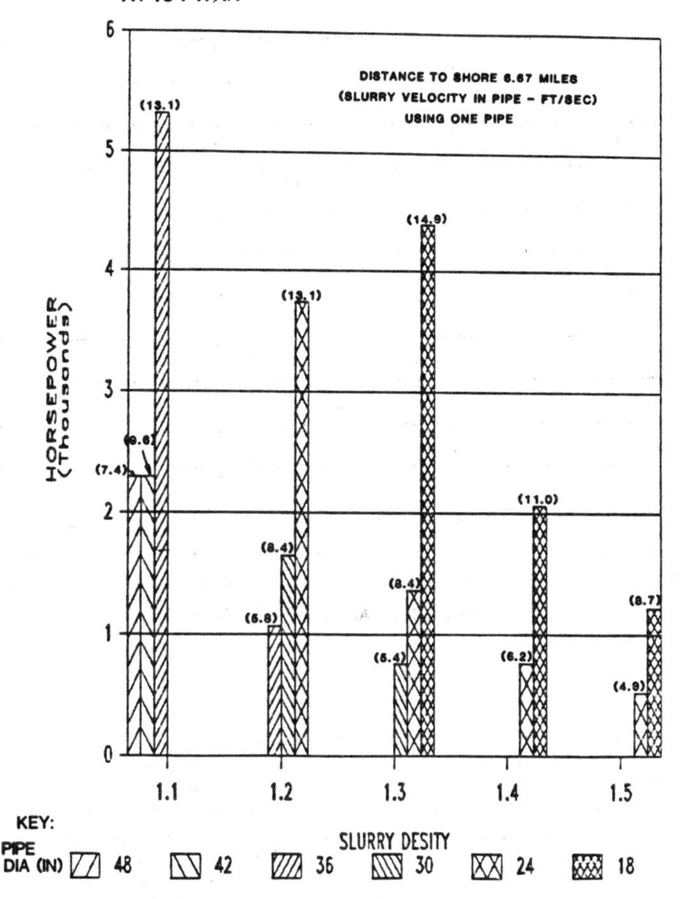

Figure 14

which would be required. Figure 14 shows the power required to pump the slurry ashore. In this case, we have included the velocity of flow in the pipe in parentheses on the top of the bar. This is very critical as we pump the slurry to make sure we are not getting the velocity too slow or the material will settle out of solution. Carbon dioxide will be produced by workers within the confines of the tunnel. We must "scrub out" this carbon dioxide in the air. Figure 15 assumes 100 people are working in the space supported by one caisson. The equilibrium levels that CO_2 scrubbers of various size can achieve are shown. These levels are by far below the ones we live with in submarines daily, 70 days at a time.

In conclusion, I would like to leave you with the following points. We believe the use of caissons can result in a significant decrease in the time it takes to excavate an undersea tunnel. Additionally, if you place the caissons above the low points of the tunnel profile, the muck handling can be greatly simplified by use of a flume cut in the invert of one or all of the tunnels. Use of the flume could achieve a major reduction in

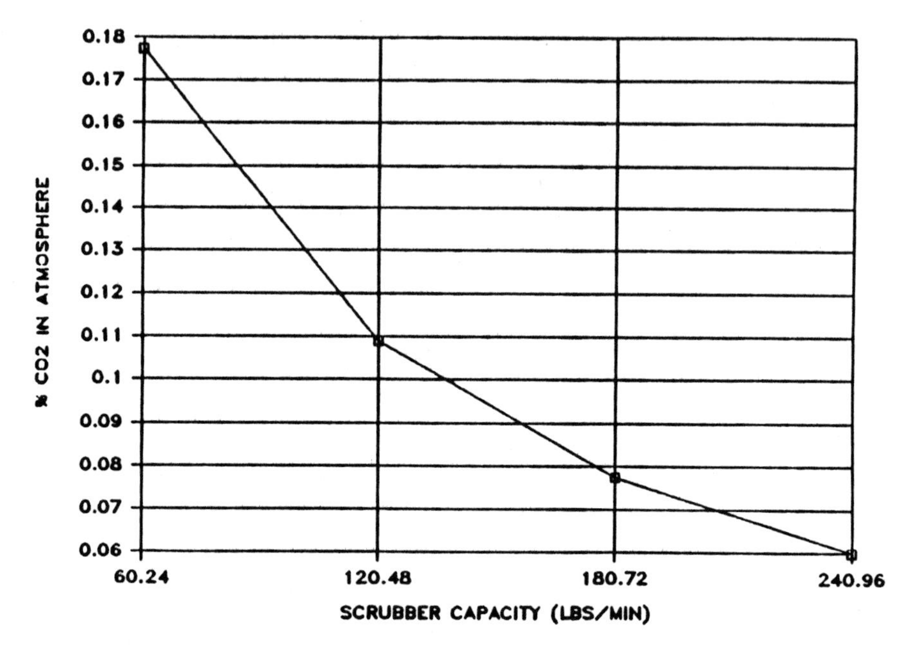

Figure 15

vehicular traffic and manpower required in the tunnel spaces. For all the above reasons, we believe that the use of caissons could result in significant reduction of cost and time. If the caisson location and design are considered in the tunnel design and mucking plans, we can actually achieve minimum interference to the surface waterways while maximizing the benefits of reduced costs and time. We feel that submerged caissons should be considered and applied to any construction effort involving tunneling under bodies of water. This is especially true when the underwater portion is of significant distance. Submerged caissons can also be used for reducing construction time when we start tunnels at underwater sites such as a bridge tunnel complex. In such a case it would be possible to be doing your underwater tunneling using the caissons while you are building your bridges and your islands, using the muck to help you build the islands.

PART III

POLICY OPTIONS AND ECONOMIC CONSTRAINTS

Some Future Directions for Underground and Undersea Transit

Introductory Remarks and Presentation of Dr. Harrison H. Schmitt

DAVID S. SAXON
Chairman of the Corporation
Massachusetts Institute of Technology
Cambridge, MA 02139

I am very pleased that you invited me to participate in this conference on tunneling and underground transport. It is a special pleasure to observe how broad the range of interests and professions is--civil engineers, economists, lawyers, bankers, academics, members of governments, and others are represented here tonight. I also note the presence of eminent authorities from across the world who have travelled far to be with us. Among them, Sir Alan Muir Wood from London; Professor Nakagawa, Chairman of the Japan Institute for Macro-Engineering; Sir Robert G.A. Jackson, the Australian statesman who still serves humanity as a principal troubleshooter of the United Nations; Arthur Bailey from Ottawa, and, of course, many others as well. On behalf of the MIT Corporation and the MIT administration, I would like to express appreciation for your attendance and for your participation. I would also like to express appreciation to Arthur D. Little, Inc., Stone and Webster Engineering Corporation, and the American Society for Macro-Engineering for their collaboration in the conference and in making the arrangements for it.

The theme of your program has emphasized that recent advances in technology have created macro-engineering possibilities in tunneling and underground transport on so large a scale that very few companies can undertake them alone. That thematic statement, which you will recognize as directly quoted from your program, of course, underscores the importance of the broad range of representation that I have just mentioned, because the kind of large-scale construction you were discussing requires the wise balancing and integration of information and expertise from an array of disciplines from both private and public sources. It also requires a marshalling of will. The great tunnels of the world and related large works constantly challenge and press forward the underlying technology and the limits of our capacity: the limits of our capacity to organize and to manage all the elements necessary for spectacular achievement. Throughout history, man's large-scale engineering projects from the aqueduct and roads of ancient Rome, the dikes of the Dutch, to the U. S. space program, have not only advanced our civilization materially but have given a special lift to the human spirit.

Greeting you and bidding you a warm welcome, I wish the conference every success in accelerating the transfer of information and of vision which is so essential to the progress of large-scale engineering. In introducing Dr. Harrison Schmitt, I need hardly remind you that as a professional geologist, an astronaut who piloted the lunar module for humanity's first venture on the moon, and more recently, as a United States Senator from New Mexico, the author of the next paper has a rare combination of credentials.

Published 1987 by Elsevier Science Publishing Co., Inc.
Tunneling and Underground Transport: Future Developments in Technology, Economics, and Policy
F.P. Davidson, Editor

Tunneling the Solar System

HARRISON H. SCHMITT
Technology and Management Consultant
Albuquerque, NM 87198

It is a pleasure to speak with you about space, about tunneling the solar system. It is my firm belief that the frontier of space--this new ocean of exploration, of commerce, and of human achievement--has produced a level of excitement and motivation among the young generations of the world that has not been seen for nearly a century.

History clearly shows us that nothing motivates the young in spirit like a frontier. The exploration and settlement of the space frontier is going to occupy the creative thoughts and energies of major portions of human generations for the indefinite future. I find it totally unacceptable and totally unrealistic that it will be 50 to 100 years before we reach major new historic milestones in space, such as that now represented by Apollo. People are not going to wait that long. Young people, in particular, will not wait that long. The only principal historical issues in doubt are the roles that will be played by free men and women, and how their roles will relate to the problems of the human condition here on Earth.

The return of Americans and their partners to deep space must be viewed in the context of the free world's overall perception of its role in the future of humankind. That role in space has not been fully formulated into a national consensus. It seems safe to say, however, that for the generations now alive, we, the United States, will be the free world's principal agent and advocate in space. As such, we must be a guiding player in interweaving the demands and opportunities of the on-going Age of Information, the soon-to-be earth-orbital civilization of space stations and lunar bases and, in the not too distant future--certainly within the lives of young people now alive somewhere on Earth--the exploration and settlement of Mars. It is my hope, as we see the third millennium turn here on Earth (it is less than 15 years away), that the effort to place a permanent human outpost on Mars will become the world's millennium project, a project I would call, "Mars 2000." (Back in 1978, I called this idea "The Chronicle's Plan," but there are still too many people who are not familiar with Ray Bradbury's remarkable tales.)

As we view this future before us, we must, first of all, recognize the obvious. Space provides the foundation for the Age of Information. Satellites provide most of the essential links in the gathering and transmission of information on a worldwide basis. Weather forecasting, resource identification and monitoring, information systems and general telecommunication systems, satellites and transocean cables are creating and expanding a central nervous system for the planet Earth. These technologies are increasingly being reinforced by the advent of modern automatic data processing. This two-way, interactive communication system for the Earth not only provides a means for dramatically increasing the well-being of all people, particularly those in the less developed countries, but also provides a means for keeping the peace. Maybe the earliest, most dramatic illustration of this central nervous system that ties hundreds of millions of people together instantaneously was seen at Christmas time in 1968. Suddenly, with the mission of Apollo 8, those hundreds of millions of people were aware of a change in a familiar object in the night sky--the moon. And with that simultaneous awareness, the world changed. It will never be the same.

Tunneling and Underground Transport: Future Developments in Technology, Economics, and Policy
F.P. Davidson, Editor

The unique resources of near-earth space provide the basis for the creation of an earth-orbital civilization. The weightlessness of space, the high vacuum at high pumping rates, and the unique view of the Earth, Sun, and stars can be utilized for industrial, public service, educational, research, and, of course, peace-keeping purposes. These new "resources" are not only accessible to Americans but to all the free peoples of the world. Indeed, they are accessible to the entire world if we are willing to act as their sponsor.

The moon, planets, and deep space offer both challenges and opportunities to excite generations of free men and women just as the New World offered both challenges and opportunities to past generations. The extension of our civilization of freedom to the planetary shores of this new ocean of space should be our basic rationale for the world's millennium project. With the successful completion of this project, namely the establishment of a permanent human outpost on Mars by 2010, we should see the first firm steps toward permanent settlements away from Earth.

The tangible beginnings of the creation of an earth-orbital civilization came with the launch of Skylab in 1974. The Skylab missions, followed by those of Salyut, Spacelab, and the Space Shuttle, began the examination of many of the potential uses of the resources of near-earth space. Skylab gave direction to our imagination of the use of these resources; Salyut, Spacelab, and the Space Shuttle give license to our imagination. Space stations, of course, will give reality to this imagination.

As I indicated, the resources of near-earth and the civilization that can be created there are basically three in number: (1) an instantaneous and continuous view of the Earth, the Sun, and the stars; (2) an infinite quantity of clean, ultra-high vacuum at high pumping rates; and, perhaps most importantly for the time being, (3) a weightless environment, that is, an environment free of most gravitational stress.

The continuous absence of gravitational stress provides a unique experimental and practical environment heretofore unavailable to mankind. Extensive experimentation and manufacturing have never before been possible where convection does not exist, even at the micro scale, where containers for fluids are not required, and where gravitationally unconstrained crystal or biological growth can occur. The absence of gravitational stress means that no vessels are required to hold experimental materials. Thus, new investigations of the dynamics and properties of fluids and emulsions, and of materials formed from them, are now feasible. These investigations, along with other basic research in physics, chemistry, and biology, are leading to many potential commercial applications of space manufacturing facilities and processes.

A return to deep space and the establishment of lunar and planetary settlements can play an extraordinarily important role in the development of an earth-orbital civilization. The potential resources that can be derived from the surface materials on the moon may well sustain both the transportation and manufacturing economies of that civilization. The principal resource that we seek to produce on the moon is, of course, oxygen. Oxygen as a fuel oxidizer and as a life-support element is tremendously important, and the moon is fundamentally an oxygen planet just as is the Earth. Principally made up of silicate, the moon also has significant resources of the mineral ilmenite, an iron-titanium oxide. A number of experimental processes, using ilmenite as a raw material, have been developed that not only produce oxygen at relatively low energy costs but also produce titanium and iron as by-products.

Relative to other resources, the principal aluminum silicate on the moon, at least in the crustal regions of the moon, is calcium-aluminum silicate. This is a feldspar mineral we geologists call anorthite. It is, of course, a potential source of aluminum and silicon. Further, as you process any surface materials on the moon, you can even extract a little bit of hydrogen. We often forget that there is a small amount of hydrogen derived from the solar wind that is embedded or absorbed in the fine particles of the moon. It is only about one cubic centimeter per gram, but any hydrogen that you can make on the moon means hydrogen you do not have to import from Earth. It is going to be, literally, worth its weight many times over in gold. Another component of the lunar surface that we sometimes forget is the material necessary to make bricks. Sintered lunar soil or sintered lunar glasses for use as radiation shielding material in earth-orbit may become very important.

The basic reason the moon is of economic interest, in contrast to what many might think at first, is that you only have to fight one-sixth of earth's gravity in order to get materials off it and into low earth orbits. That is much less costly once the mining and processing plant is established than bringing the same materials from the surface of the Earth into the same low orbits.

No matter what other justifications may be given, the ultimate rationale for today's generation to return to deep space, to the moon, and elsewhere, and to establish a permanent presence there, is to create the technical and institutional basis for the settlement of the solar system and, most urgently, the settlement of Mars. I am not saying this off the top of my head, ladies and gentlemen. I have spent a lot of time through the last 15 to 20 years with the young people of this country. When you ask them a series of questions, you find out where their hearts and minds are. I have done the same throughout the world in my travels as an astronaut.

When you ask young people, "How many of you would like to go to the moon someday?" you get about 75% of the hands in a given class. There are inhibitions when you get to the level of high school, but those can be broken down with a little conversation. But when you ask, "How many of you would like to go to Mars someday?" you get about 85% of the hands. Why the difference? I have asked them, and they say, "You've already been to the moon." It is this kind of attitude that is setting the stage for the first great adventure for humankind in the third millennium after the birth of Christ.

Steadily increasing philosophical and psychological momentum for this adventure on Mars is "a-building" among the young people of Earth. Rather than go to the moon, they have their eyes on the next frontier beyond the moon. They are the ones, those 10% who have lost interest in the moon, who will go to Mars one of these days--soon. They are the ones, like most of our ancestors before them, who will never be satisfied with either the comfort or the restrictions of home and Earth. They are the parents of the first Martians. They are in our homes and in our schools, driving us to distraction as they struggle toward adulthood. They have no technological barriers nor, since Apollo, any psychological barriers between them and their migration into space.

There should be no economic barriers to the settlement of the solar system. We did not go to the moon because we spent $22 billion over ten years or, in today's dollars, $75 billion. We got there because young men and women believed that this was the most important thing they were going to do with their lives. It may turn out not to have been that important, but they believed it was at the time. Personally, I believe it was the most important thing that 500,000 Americans did with their lives in the 1960s and 1970s. When young men and women believe this way, whether they are American or European or Chinese or Soviet or whatever, then anything is possible. It is possible that some of these believers will, indeed, become those parents of the first Martians.

The importance to the parents of the first Martians of a self-sustaining settlement on the moon, trading directly with an earth-orbital civilization of permanent space stations, is that it provides the technical and institutional basis to go to Mars with the purpose of establishing a permanent outpost with the first few expeditions. These expeditions could be initiated by the end of the first decade of the third millennium from the technical foundation of space stations and a lunar base. A permanent settlement on Mars will take a little longer, but a permanently occupied outpost, resupplied by a permanent system of interplanetary space stations, clearly will be technically possible as well as philosophically desirable soon after the establishment of a permanent lunar settlement, if not before.

Now, I hope that you will not think that I am just playing to the attendees of this conference when I say that the science, engineering, and art of tunneling will be one of the most critical technologies in humankind's future in space. "Tunneling the solar system" is not just a catchy, throw-a-way title. In fact, several of the speakers at this conference this afternoon made positive reference to such activity.

First of all, as we settle the moon and Mars, tunnels and tunnel-like covered surface excavations will be our first line of protection against the radiation dangers of sporadic solar flares and ever-present cosmic rays. This will be particularly true on other moons and planets with little or no protective atmosphere or magnetic sheath. A few meters of rock or consolidated rock debris will be enough for such protection around habitation facilities and fixed work sites to make long-term or permanent activities possible.

The thick cover of impact-generated debris (regolith) on the surface of the moon will probably lend itself to explosive excavation of near-surface tunnels, or "trench-tunnels" if you wish to call them that, into which prefabricated or inflatable structures could be placed and then covered by the excavated debris mixed with an adhesive. As most mining operations currently envisioned for the early moon bases will involve extraction of ilmenite, native iron, hydrogen, and aluminum silicates from the already finely pulverized portion of the surface debris, tunneling into lunar mountains probably will not be required in the foreseeable future. This assessment, of course, could change very quickly in the unlikely event that water permafrost is discovered in the permanently shadowed walls of polar craters and valleys.

The known existence of water on Mars makes it probable that tunneling will be used there in water-ice mines. Tunneling also may be needed to extract various ores from concentrated mineral deposits, not unlike some deposits with which we are familiar here on Earth.

If it is proven that one or both of the moons of Mars contain significant amounts of water-ice, as current theory suggests, tunneling also may prove to be the most efficient way to mine the ice as a first step in creating a spacecraft refueling station in Martian orbit. Similarly, mining the resources of captured earth-crossing asteroids, should that ever prove feasible, will most likely be done through tunneling activities. All such tunneling also will create habitation space.

Let us look more at the early Mars missions. As I indicated earlier, the first Martian base will probably be established by the excavation of trenches or tunnels in which prefabricated or inflatable shelters can be placed. The most likely location for a base, I would think, would be in one of the deep equatorial valleys near areas that show strong photographic evidence of being underlain by water-rich permafrost. The deep valleys also will provide somewhat higher temperatures and atmospheric pressure than other possible sites, such as the polar regions where we know water-ice is present as ice caps. However, one of the advantages of using a large interplanetary space station for the first trip to Mars is that the time necessary to examine the Martian moons for resources and to select a proper site for the first base can be spent in orbit before committing to the first landing. If necessary, reconnaissance of several sites can be carried out before committing to a site for the base. Unlike the Apollo program and our current Space Shuttle program, mission control will almost certainly be in such an orbital station rather than on Earth.

The most inconvenient aspect of living and working on Mars will be the dust. For those of you young enough now to contemplate going there, it is going to be dusty every once in awhile. Unlike the moon, which has no atmosphere, fine dust on Mars blows around in great global storms and settles very slowly over a period of months. It may be that these storms will require the temporary confinement of the explorers to shelters, much like in winter at Antarctic bases. It also may be that it will not require that; we just do not know that much about it. It is probably very fine clay that is blowing around. It may also be that the wind has sorted out many useful minerals on the basis of density. Minerals such as the ilmenite that I mentioned earlier could be extracted from the concentrated deposits. Who knows? Maybe there is gold-flour lying around. That ought to get everybody excited.

The major uncertainties of Mars, and particularly for the first project to go to Mars, are the chemical nature and agricultural potential of the Martian soil. It is highly probable that, like the lunar soil and new terrestrial volcanic ash, the Martian soil is very fertile. At least there should be fertile areas even if many soils are unusually oxidizing as some data suggests. Soils may be rich in particles of clay as well as impact and volcanic glass, all of which will help to make them quite fertile. The existence of a Martian atmosphere, its oxidizing nature, and the possible presence of sulfur compounds in the soil crusts may mean that some chemical treatment of soils will be required before they can be usefully farmed. The alternative to traditional means of growing food may be hydroponics. This option would depend on the local availability of sufficient quantities of water in the vicinity of local farming operations. Clearly, the early interplanetary space stations used for transport to Mars must be large enough and capable enough to be outfitted for the exercise of either of these options. Large quantities of imported food will be required as well until reliable agricultural production is established.

Unlike the moon, Mars may have had an evolutionary phase not unlike that of the early Earth which included significant water erosion. The details of why that is possible are beyond the scope of this paper, but if it happened, then several important consequences follow. First of all, we will be dealing with stratified sedimentary rocks in some places on Mars in addition to the volcanic and impact debris that we know exists.

Second, we may well be dealing with the relic effects of biological activity. As you may know, we have not yet discovered any direct indications of existing life on Mars, nor do I expect that we will. However, the early phases of history on earth and possibly on Mars ncluded a reducing global sphere of fluid. Conditions in such a fluid phere on earth were conducive to the development of very simple forms of ife through a very complex but largely unknown process. Not only have we nown for a long time that various algal species formed within the first illion years of earth history (4.6 billion years total), but now, as a esult of discoveries in Australia and southern Africa, we know that there ere bacteria at least 3.5 billion years ago. Algae and bacteria are volutionarily complex enough that much precursor activity had to go on in he preceeding billion years. The same evolutionary sequences may have aken place on Mars. If there was early biological activity on Mars, not nly would we expect to find the fossil record of that activity, but we ight expect to find hydrocarbons, particularly methane. It may even be hat biological activity on Mars was able to progress far enough to result n an oxygen atmosphere for a limited period of time. That oxygen tmosphere appears to have appeared on Earth only about two billion years go, maybe even a little bit later. Obviously, Mars evolved differently rom the Earth after that time for now there is very little oxygen in the artian atmosphere. A great deal of oxygen, apparently, is locked up in he Martian soils.

It is doubtful that significant biological forms exist today. If any o exist, there are only a few environmental niches where they might be iable. We have not visited those niches with the Viking Landers, but here are some that explorers might target for on future missions. Recent iscoveries of biological niches and survival strategies related to the arsh environments of Antarctica and the deep sea once again illustrate hat we should not underestimate nature. Nevertheless, life as we know it ould not be sustained by the Martian planet without our intervention.

I have pursued some of these issues in reasonable detail because I ant all of the participants at this conference to begin to think about he kind of macroprojects that exist on planets other than the earth. learly, there are many important macroprojects for our earth, but as we ollow the eyes of the young, we can see that there is even more to be one elsewhere. However, in spite of all the technology that we have, ncluding that which you have been discussing at this conference which is mmediately appropriate to the questions I have raised, space activities f the future will be sustained less by technology or by knowledge than it ill be sustained by emotions--the emotions of the young men and women of he planet Earth who wish to go there. Indeed, young Soviets, East uropeans, Chinese, and Cubans who do not feel the freedom that we feel ere tonight must also look to space as the earth's frontier as we do. As ith our ancestors, their freedom lies across a new ocean, the new ocean f space. The millennium project that we should consider, Mars 2000, is, believe, their hope as well as our mission.

Planetran Today: Prospects for the Supersonic Subway

ROBERT SALTER
Litton Industries
Beverly Hills, CA 90213

FOREWORD

This report is dedicated to those brave souls who strive to bring back the golden era of railroading. It is hoped that something like Planetran will help in railroad resurrection--providing expeditious service but simultaneously providing for the revitalization of existing local rail links and permitting their leisure utilization.

Let us pray that Willie Nelson's "City of New Orleans" does not indeed become a forgotten native son.

From one who has ridden the "Sunshine Special" from Los Angeles to San Antonio, the TransSiberian railroad from Nahodka to Kabarovsk, and has been a "straphanger" on the Shinkansen "Bullet" train, I fervently hope that these denizens of the past will continue in some similar form into our future world.

INTRODUCTION

My chapter deals with ultraspeed, electromagnetically propelled and levitated transportation systems of the future. Such systems should be able to carry passengers across the United States in less than an hour in a quiet, economical, fuel-conservative, and nonpolluting manner. They would be designed to optimally interface with local automotive and mass transit systems. An initial candidate for future transportation is called "Planetran," whose ultraspeed trains float on magnetic fields and travel in evacuated tubes in a system of underground tunnels.

I will first make some remarks on transportation in general followed by background, concept description, critical issues, economics, related technology, other applications, and summary.

GENERAL

Transportation is, of course, a major segment of our society--it affects all of us, every day. It is an area that has grown classically like Topsy; no particular advance planning and very little in the way of institutionalized research.

Fortunately, transportation is a field rich in innovative opportunity and so it has never lacked for technology. However, it has perennially suffered from a lack of management and planning. A good example is in the magnetic levitation (MAGLEV) area. Electrified rail systems were well developed in the 19th century. Also present in that era were numerous developments (and patents) on magnetically levitated devices. The MAGLEV vehicle concept was patented in 1912. Why has it taken 70 years just to initiate MAGLEV research? What do we know today that we did not know in 1912?

Part of the answer is in overall transportation development trends. Before trains and motorized river transport, transportation was of a dispersed nature. Trains changed that to a great extent in the long-haul transportation sector. Society became "corridorized." Cities grew and

Tunneling and Underground Transport: Future Developments in Technology, Economics, and Policy
F.P. Davidson, Editor

rospered along major rail links. Large cities installed electrified
ocal transit with street cars and interurban trains to interface with
nese links.

With the emergence of the automobile and airplane early in the
wentieth century, the pendulum swung back again. The flexibility of the
utomobile allowed our society to grow in a diversified fashion. The
irplane was freed from the long-link corridor constraints of the railroad
but initially introduced similar constraints of its own).

In recent years with increasing urban congestion, remoteness of
irports, the energy crunch, and the mythical 55 m.p.h. limit, interest in
ass transit and longer-haul train systems has re-emerged. Initially, it
as fashionable to espouse the efficacies of such approaches as part of an
verall conservation/environmental cult. The automobile became the "bad
uy," destined to be replaced by other methods; this notwithstanding the
act that automobile travel had represented over 90% of <u>intercity</u>
ransport for four decades or more.

During the early 1970s, the Department of Transportation (DOT) was
oing some research in advanced transport systems. Their Office of High
peed Ground Transport (OHSGT) sponsored several historic MAGLEV
rojects. Their application studies were 100% devoted to rail transit
hich in turn was quite out of tune with the times. If one had the
isfortune to arrive in, say, Philadelphia on the train, there was little
vailable in the way of infrastructure at the terminal--such as a rental
ar agency.

In that era I spent some time trying to sell the DOT on sponsoring
esearch on the VHST/Planetran concept. On one visit the official in
harge received me politely but pointed out that they were not even going
o continue the OHSGT's near-term research (it was destined to take on
MTRAK-revival support studies). Also, he personally had little time for
iscussion--he had to leave early--he was <u>driving</u> his family to St. Louis!

The automobile has not only survived, it is burgeoning in other parts
f the world. It has invaded Japan--a country that already had a superbly
ntegrated rail system. Europe and Great Britain have taken the lead in
igh-speed auto travel. Their freeways operate effectively and safely at
wice the allowable speed on U.S. highways--despite their more limited
uel resources.

In fact the automobile is becoming a way of life throughout the
orld. Last December my wife and I were caught in the early morning
ommuter jam on a multi-lane freeway--smog and all. The unusual part is
hat it was on one of the Canary Islands on the Atlantic off the coast of
frica!

The moral of the above is that the automobile offers unparalleled
lexibility in travel and is endemic to our living and demographic
atterns. It will be with us for years to come. However, the IIAS, an
nternational think tank in Laxenburg, Austria (U.S. sponsor is the
ational Academy of Sciences) forecasts that ground transport systems of
he 22nd century will be <u>all</u> electromagnetically driven in view of
fficiency and energy conservation constraints.

In the nearer term, we must consider the automobile and existing
etropolitan transit systems as fundamental building blocks in concert
with which a system like Planetran will be developed. The airplane, with
its deficiencies and attributes, must also be considered as a factor.

As we will see later, Planetran is a system that must be fully developed before it will be effective. Its development will represent a very bold step. There are ways to mitigate this problem such as developing intermediate uses of its tunnel system. The dominant problem is initial expense. It does not have the auto's initial flexibility of use but at the time of Henry Ford, if we had to arrange for our present $500 billion freeway system, automobile development would have been badly stifled.

There is one other aspect of Planetran that we must consider. Even if it offers economy, conservation, and pollution-free operation, people will not ride it unless it also offers convenience--convenience both in speed of operation and in ease of access.

BACKGROUND

My first consideration of the Planetran concept occurred in the mid-1950s. The Air Force had been given custody of the U.S. satellite program in 1949 and they, in turn, initially vested it at Rand; later at Lockheed (where the Air Force Space Program now still largely resides). I was the Project Director both at Rand and, in the early phases, at Lockheed. One of our tasks was to assure that future studies on scientific and domestic uses of space were assured. In the course of this work we examined the prospect for ballistic vehicle application to public transport.

It was patently obvious that it was more efficacious to create outer space conditions on or under the earth's surface than to attempt to put transit systems into outer space. We needed to operate vehicles at ultra-high speeds which meant evacuated tubes. Electomagnetic levitation was an obvious choice and superconducting magnets were a well-known concept--if not development--at that time. Thus, the initial conception of Planetran did not require a burst of genius. It did, however, require some clear thinking on aspects of the overall system.

About that same time another member of Lockheed left to start a company to develop a gravity-driven subway system. Such a system has some merit for local transit but it is not a candidate for the long-link systems which are the key to Planetran and any effectively integrated overall transit approach.

The problem is that the real cost of the system is the tunnels--over 90% of the cost. We cannot afford to operate at slow speeds in such a system. Indeed, even if we were to conjure up a very expensive ultraspeed approach using aircraft-type vehicle bodies, superconducting magnets in the vehicles, completely active electromagnetic guideways, etc., the tunnel cost would still predominate--it would probably still be 90% of the cost or more. We cannot afford to go slow--not only for system economics but for rider acceptance.

Once we have established this precept we see that the only limitation to operating an electromagnetic system, not only at aircraft speeds but at much higher velocities, is one of control. Tunnels are needed with very large radii of curvature. Sadly, the last constraint appeared impossible in the late 1950s--the control problem was unsurmountable with the then extant computer technology.

A decade later, when I was at Rand, I was asked to suggest several future developments that Rand might pursue. At that time microelectronics was beginning to evolve and we could foresee the future prospect of microprocessors and minicomputers.

We tried to patent some of the innovative aspects of my revised oncepts. Among these were programmed gimballing of the passenger seats o align them with the resultant of thrust/braking forces and gravity. nother basic concept, not previously considered, was an outgrowth of the xtremely demanding vehicle control necessary to avoid deleterious ransverse acceleration on passengers when the train was traveling at everal thousand miles per hour. This system not only considered complete omputerized microcorrections at many stations along the guideway, but adio frequency/optical sensing of minute excursions in vehicle attitude nd/or velocity vector values which would be sent on ahead at the peed-of-light for subsequent course correction. Further, we provided for ny excursion in the tunnel and/or the evacuated tube (such as from earth remor, etc.) by placing inertial detectors on such structures, and odifying Planetran's electromagnetic path accordingly. This was ertainly a unique and new concept. In essence, we could stabilize lanetran's electromagnetic path in inertial space with this microcomputer ystem.

Unfortunately, this computerized system, which never before had been onsidered (and probably nowhere else since), was not favorably approved y the patent examiner. Train systems do have computers--they issue ickets. Vehicles, including aircraft, do have inertial guidance ackages. Rand, not being a profit-oriented company, did not have the esources or incentive to appeal the examiner's misconceived decision on his point.

However, in the case of the gimballed seat concept, we were in for nother surprise! Such a seat had been patented in 1945 by Dr. Robert oddard, pioneer of U.S. space research. However, there were no other atents that would give a clue as to what kind of system needed these imballed seats.

Further probing revealed that Goddard had indeed invented Planetran. is concept involved employing electromagnetically suspended and driven ehicles in evacuated tunnels, and employing superconducting magnets in he cars to increase levitation gaps to about a foot for viable control spects. He postulated going from New York to Boston in ten minutes. But e also struck out with the patent examiner. Even the appeal, which ncluded the exhortations of Dr. Aubrey Austin, Director of the U.S. cademy of Sciences, was to no avail. (One of the vagaries of U.S. patent rocedures is that the examiner's findings are generally considered quite nviolate.) The examiner was right in his decision--but for the wrong easons. Goddard's principles of locomotion and support were correct--but e had not solved the microcontrol problem.

My talk today deals with ultraspeed, electromagnetic transportation ystems. Such systems are both levitated and propelled by powerful lectromagnetic fields. With proper design, ultimate speed is limited nly by materials and structures. As an example on a much smaller scale, ail-guns in the Department of Defense (DOD) laboratories have already chieved speeds of over six miles per second--30% faster than earth atellites (and in some of these experiments with a gun only 6 feet ong!). Near-term weapon programs expect to quadruple this speed. Impact hermonuclear fusion applications forecast future pellet speeds of 100-200 m/sec.

Japan National Railways (JNR) several years ago demonstrated speeds of 325 m.p.h. with a MAGLEV experimental vehicle at their 7-kilometer-long est track at Miyazaki, Kyushu. A 20-ton vehicle traveling at this speed as a kinetic energy (KE) of 390 megajoules (MJ). If it is accelerated at a quarter of a "g" (about that employed by JNR at Miyazaki), then about 7 megawatts (MW) of kinetic power must be received by the vehicle.

By comparison a typical weapon rail-gun test bed will accelerate a 2/3 pound projectile to 4 km/sec. The projectile's KE is 2.5 MJ. However, it reaches this speed in several milliseconds so that received pulse power required is roughly 1000 megawatts.

A number of major corporations (as well as national and university laboratories) are developing electric guns for both tactical and strategic weapons (including SDI space-based hypervelocity guns), impact fusion, and air vehicle launch. These companies include Boeing, General Electric, Westinghouse, Ford, General Dynamics, Hughes, etc.

NASA has studied electric launch for space vehicles. This was proposed to the Air Force Space Division by the author in early 1962. The proposal called for a traveling wave type accelerator driving a 1000 pound projectile containing its own dipole field from an onboard power supply. Launch at 30 degrees elevation angle and 40,000 ft/sec (12 km/sec) was assumed, launching a number of successive satelloids for "once over" type reconnaissance.

Figure 1 is a chart that illustrates various electrically accelerated devices; it displays the relative regime in which Planetran operates.

Figure 1. The kinetic energy of hypervelocity macroparticles as a function of particle mass and velocity for a range of masses and velocities of interest.

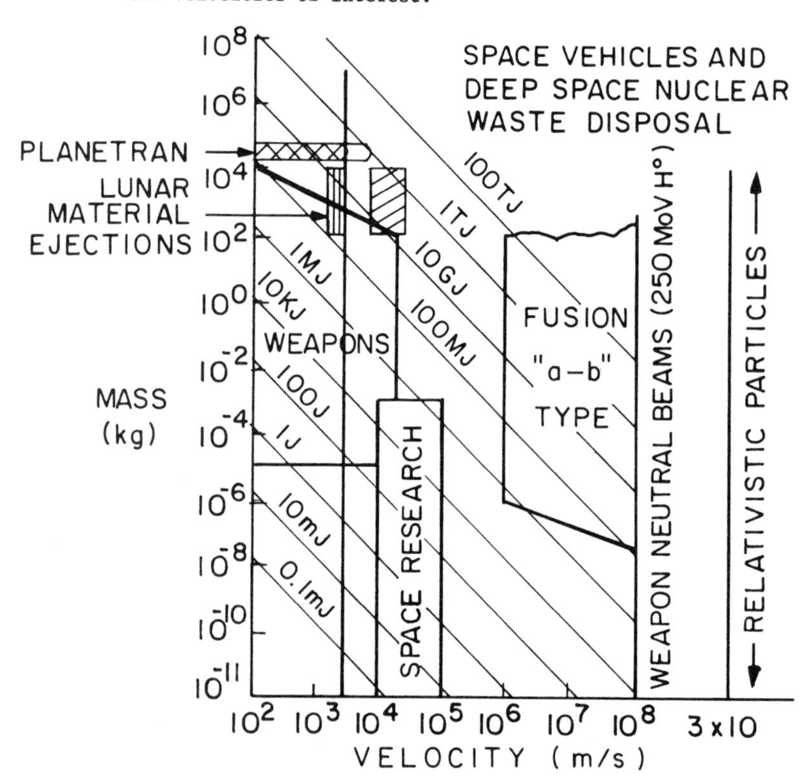

NCEPT DESCRIPTION

The Planetran concept has been well documented over the past few years
numerous articles, books, and periodicals. The February 1978 Rand
aper P-6092 (essentially the talk given to the AAAS in Washington D.C.
pring: 1978 Meeting) is still "au courant." This same material is
ncompassed as an integral component to various significant treatises on
macroprojects" by Professor Frank Davidson of the Massachusetts Institute
f Technology.

As mentioned previously, Planetran is a form of "MAGLEV" vehicle that
erives not only its support from electromagnetic forces, but also its
ropulsion/braking forces. In this respect, it rides and is propelled by
agnetic waves in a fashion similar to a surfboard (see Figure 2).

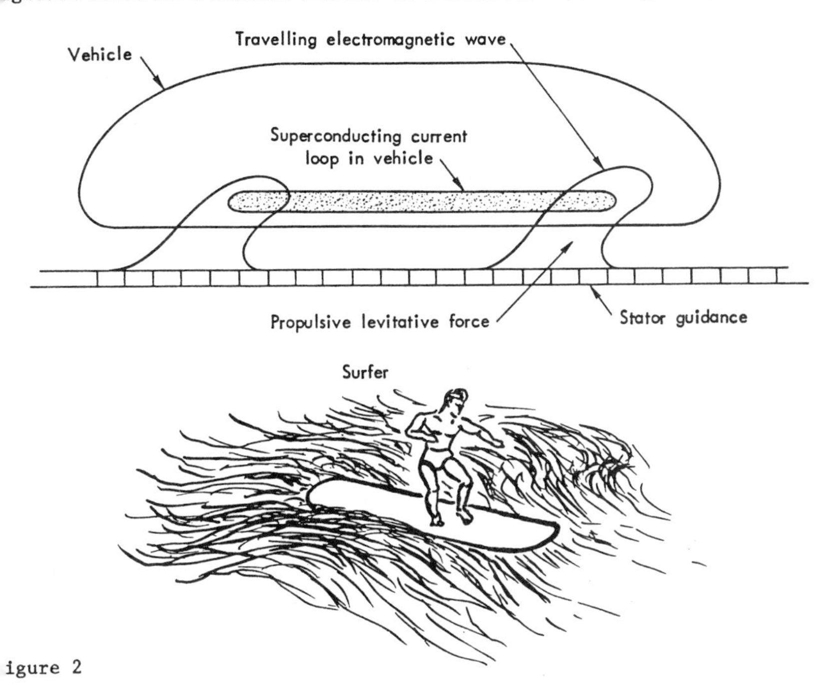

igure 2

In my first analysis of the Planetran Concept I assumed a nonstop,
oast-to-coast link with acceleration/deceleration forces of one "g." A
rip from Los Angeles to New York takes 21 minutes. At a 1 "g"
cceleration, with seats that are programmed into optimal positions, the
assengers feel about 40% heavier at the outset and the finish of the trip
ut, due to centripetal forces, feel very nearly their own weight at the
idpoint.

In order to provide more universal service, and also to reduce
assenger accelerations as much as possible, we have assumed that
lanetran stops at Dallas along the way from Los Angeles to New York.
ars either pause and then continue or else are rapidly replaced by an
lternate car. One-stop, through passengers take 54 minutes to get across
he country-- more than the previous 21 minute case, partly because of the
ne stop, but also because we have reduced acceleration/deceleration from
constant 1 "g" to a constant 1/3 "g." At this latter thrust level (and

with programmed seats), the passenger only feels 5% heavier throughout the trip. We doubt that passengers would be able to detect this 5% weight difference. Figure 3 shows main Planetran cross-country corridors as well as shorter "feeder" links.

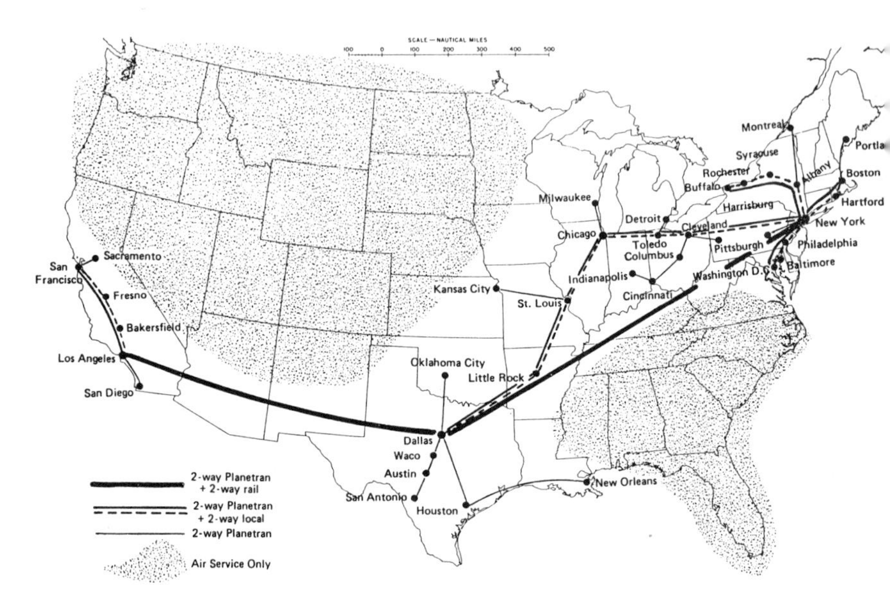

Figure 3. The early Planetran U.S. route system.

A number of MAGLEV systems or developments exist in the world. The shorter-range, lower-speed method employing attractive magnetic forces was originally pioneered by the Germans, and later taken up by Japanese Airlines (JAL) for a Tokyo to Narita airport link.

This approach gets less efficient at higher speeds. Also, it involves relatively small clearance gaps between the vehicle magnets and the "rails." The author rode the JAL prototype on a 2-kilometer test track on an island in Tokyo Harbor. This vehicle attains a maximum speed of 175 km/hr in a short distance. The author was furnished with a recorded tape of the vehicle excursions. The overall clearance gap was 1 centimeter and the servo system held maximum excursions to approximately 1 millimeter. Obviously such a small clearance does not allow much "rattleroom" and even a small obstruction may be a problem.

The alternative to the magnetic attraction approach is to employ repulsive forces. This has been championed by Laithwaite in Great Britain, by other countries including Canada, and by the Japanese National Railroad (JNR). Use of supermagnets (electromagnets with superconduction currents) in the vehicle operating against ordinary magnets in the guideway will produce high enough magnetic repulsive forces to loft the vehicle about a foot above the guideway.

JNR has been experimenting with this type of MAGLEV for a decade or more. With this type of vehicle levitation, maintenance requirements over the years have been nil. This is in contrast to the Shinkansen bullet rain where they need to use one out of three shifts each day to repair ehicles and guideways damaged by Shinkansen's high vibration and shock evels. There have also been numerous complaints from Shinkansen's eighbors regarding noise and vibration.

JNR's MAGLEV originally had a T-shape guideway but revised their 7 km-ong track to a saddle-or U-shaped guideway in recent years.

Planetran has been assumed to have a U-shape configuration from the utset. Its superconducting electromagnets operate on over a million mperes of current and Planetran does indeed "float" a foot or so above ts guideway.

There is a basic difference between Planetran and other MAGLEV schemes n that the other schemes have separate levitation and propulsion systems--n some cases propulsion is other than electric.

Planetran not only uses an integrated propulsion (or retardation)/evitation approach but further, the particular electromagnetic force arameters are precisely optimized for an exact value at each point along he way. Every car traveling over a given link is required to follow xactly the same velocity/acceleration profile. With systems precisely ailored in this way, we estimate overall electrical efficiency approaching 8.5%--a figure now achieved in large electrical generating stations mploying synchronous machines.

RITICAL ISSUES

As we have noted in the foregoing, technological development is not a ritical issue. Nor is passenger acceptance--particularly if we do not xceed 1/3 g acceleration in the system. In order for Planetran to unction properly very precise control is demanded so that the passenger ill not feel jolts, vibrations, or sideways accelerations. The car nterior should be very quiet, individual VCRs plus background music can e employed to entertain passengers for the 54 minute trancontinental unket.

Critical issues center around economic, political, and management spects. For one thing there is no jurisprudence for underground systems quivalent to "Freedom of the Seas" or "Freedom of the Skies."

Another issue involves demographics--do you put in Planetran corridors o maximize service to existing population distribution, or should they be laced such as to enhance new growth areas?

Acquisition strategy is an important issue. Planetran's success inges on development of the transcontinental links. Until these are in lace and operating the other parts of the system are not economical. One pproach details a strategy for development of Planetran that assumes nitial Department of Defense (DOD) sponsorship. Conventional railroad ines are placed in the transcontinental tunnel segments--initially to elp in tunnel construction, later to enhance the existing railroad ystem. DOD would benefit from this railroad upgrade as well as consider he tunnel utilization for strategic communication links, civil defense, nd (possibly) for space strategic missile (DUG) basing.

Subsequently, tunnel space could be rented out to clients with video/iber optic links, pipelines, transmission lines, other utility and ommodity transport, etc. At such times as economics warrant, evacuated ubes for Planetran can be placed in the tunnels by, perhaps, industrial onsortiums.

ECONOMICS

Economics plays a major role in both the design and development of Planetran.

As we have noted previously the Planetran tunnels represent 90% or so of the total system cost. Our present estimate of overall tunnel cost based upon the main links and feeder lines shown in Figure 3 is $500 billion. Spaced over a period of 10 years this is not a large fraction of our Gross National Product (GNP)--particularly in comparison with macroprojects of the past, such as the pyramids of Egypt.

Characteristically, underground rail systems have cost far more than original investors anticipated and they seldom received a decent return on their investment--but society in general was greatly benefited.

Actually with the high volume in freight and passengers, and with revenues from other tunnel users, Planetran might go a long way toward paying for itself out of revenues alone. We have made cost estimates based upon "$1 per minute in the system" passenger fares (i.e., coast-to-coast for $54) and predict $20 billion/year revenues for the high-speed portion of Planetran alone.

In order to minimize tunnel costs, we assume multi-links in each tunnel--up to four evacuated tubes. The tubes are relatively small (16-feet diameter) necessitating that the Planetran cars be long and skinny and that a number of cars would be connected together in true train fashion.

Planetran is designed to be highly efficient. As discussed above, tailoring the electrical propulsion/levitation system to a precise, single-point operation at every point in the system means that we can probably achieve the 98.5% efficiency now realized in synchronous electrical machines in present-day power stations. Thus, since Planetran cars are braked electrically, returning power back into the system, the total propulsive loss in bringing a train up to speed and then stopping it is about 3%.

Planetran's tubes are evacuated down to 0.1% of sea level air density (actually the atmospheric density at 52 km/32 miles altitude) so that atmospheric drag is virtually zero. Energy must be expended in maintaining this vacuum and is included in the economic analysis.

Planetran's form of magnetic suspension is highly efficient--active guideway electromagnets operating against vehicle-contained supermagnet fields. This drag is far less than that of a highly efficient aircraft operating at great altitudes, for example.

Planetran does not need to climb to altitude or to expend its kinetic energy as atmospheric thermal pollution as does aircraft when it returns to earth. Planetran would use about $1 of electrical energy in carrying a passenger across the USA. This electrical power can be generated in highly efficient and nonpolluting dedicated nuclear power plants.

RELATED TECHNOLOGY

In the foregoing part of this paper I described work underway in electric guns for weapons and for impact fusion. These developments, in turn, have been highly benefited by military electrical pulse-power system developments for radars, lasers, and ship propulsion, and by large electrical supplies and magnets for thermonuclear-fusion electrical power

tation research. This latter work has also pioneered liquid helium-
ooled superconducting magnets such as would be placed in the Planetran
ars.

The MAGLEV developments of Germany, Japan, and elsewhere are, of
ourse, important to the Planetran concept--particularly the work of JNR
n their MAGLEV train prototype which does include superconducting
agnet-type (repulsive) levitation.

Microprocessor/minicomputer computing systems and various types of
ensors for distance and speed measurement are now an integral part of our
ociety--particularly the applications to space technology.

THER APPLICATIONS

We have mentioned other uses for Planetran tunnels under economics.

The initial placement of conventional rail lines in Planetran tunnels
ids in its development and also upgrades the U.S. rail system. Actually
hese tunnel-located rail lines can do more then simply enhance adjacent
xisting rail lines. By employing special measures, and traveling at 100
.p.h., conventional steel wheels on rail trains could move as much as 7
illion tons of freight across the United States every 24 hours to
ontainer ships, factories, etc.

Another very unique application of a Planetran-type concept is that of
elivering materials or documents on a high-acceleration basis: time-
ritical factory or laboratory, material, or biological processing, or
imely delivery of "hard-copy" legal documents, photographs, etc.

Yet an even more unique use of Planetran is the potential
ong-distance delivery of material or personnel at faster-than-satellite
peeds. Low altitude earth satellites travel at 5 miles/second (i.e.,
000 miles in 10 minutes). The centrifugal force generated by a body
raveling at this speed is exactly balanced by the earth's gravitational
orce--hence the principle of space satellites.

If we wished to travel at twice satellite speed (i.e., cover 3000
iles in 5 minutes) it would be necessary to provide a downward force on
he body of 4 times the earth's gravity attraction ("g") of which 3 "g's"
ould need to be supplied by rockets. This would be a prohibitive rocket
equirement. We conceivably can do this with Planetran--placing the
epulsion electrical conductors in the ceiling of the tubes for the
super-satellite" phases of travel.

UMMARY

Planetran portends to bring about a new epoch in the annals of
ivilization. It could break the logjam building up in our air travel
ystems for example. (It is interesting to note that both the overall
lapsed travel time and cost to fly, say from Atlanta to Dallas, is about
he same today as it was 50 years ago.)

Planetran, with an associated freight system, could go a long way
oward improving the environment--ridding highways of trucks and much of
he long-haul auto traffic. It would greatly reduce the need for air
ravel and thereby remove a source of both hydrocarbon and thermal
ollution--let alone noise and vibration.

Because of Planetran's high efficiency additional sources of
tmospheric pollution can be reduced. Further, Planetran would not need

to combust transportable fossil fuels for its propulsion--clean, abundant fission and fusion electrical power plants would be employed.

Planetran's tunnels offer a safe, benign, and economic conduit for various communication and utility transmission--again reducing energy waste (and inherent pollution); but also removing the blight of surface pipelines and overhead transmission wires.

Planetran's development must be highly structured--it must not be allowed to grow like Topsy. However, its evolution must follow reasonable laws of supply and demand with incentive rather than subsidy being the mechanism by which it is created.

With ingenuity and political finesse it should be possible to build a world-wide Planetran system creating even more of a one-world situation than did the jet airplane and its predecessor the clipper ship.

On the other side of the coin is the rampant governmental bureaucracy which is ever increasing at exponential growth rates throughout the U.S. and the world. Bureaucracy has already stultified many countries of the world. Its force in the Soviet Union is probably our only real protection from the Russian Bear's military and hierarchy.

A decade ago the Rand-New York Institute (a joint venture between Rand Corportion and the City of New York) tried solving some of the city's socioeconomic problems. In the arena of housing and rent control alone they found over 150 different overlapping agencies that had to be dealt with.

How then can a macroproject of the magnitude of Planetran be deployed? There will, of course, have to be rulings regarding a "Freedom of the Underground" policy, some arrangement for compensation or licensing of oil and mineral deposits discovered in Planetran tunneling projects, and use of "Eminent Domain" in many instances. Local support for Planetran may be gained through creation of, say, 100 25-mile-long tunnel projects for a 2500-mile link. The overall project must enjoy the status of a national critical objective such as the Apollo program and it will also require the discipline, big systems procurement procedures, and systems engineering of DOD-managed projects.

Is Planetran too far-fetched to be a reality? If we look back a century or so we would guess that our present jumbo jets would be a wonderment let alone that of man walking on the moon. Actually, most of the concepts of Planetran were known at the turn of the century. Planetran's pursuit may actually suffer because it will not require technical and scientific breakthroughs and therefore will not be enough of a challenge.

Planetran will be, however, a prodigious undertaking and its overall contribution to society must be fully understood before support for its creation can be gained.

ıfluencing the Cost

:RBERT H. EINSTEIN
'ofessor of Civil Engineering
ıssachusetts Institute of Technology
ımbridge, MA 02139

ITRODUCTION

A summary only of the talk presented at the Symposium will be given ıre. The material relating to productivity and technological factors :fecting cost will be written in a paper which is, at the present time, ı preparation. Aspects involving the uncertainty and thus the risk ıctor in tunnel economics have been extensively treated before (see e.g., ınstein et al. 1974, Ashley et al. 1981, and Ioannou 1984).

As with all construction projects and in particular with large :ojects, not only construction cost but construction time through its :fect on return on investment have to be considered. The latter issue ıll be discussed in more detail by Professor Thurow. Nevertheless, in ıe following lines the possibilities of influencing construction cost and ıme will be discussed. Time and cost are influenced by productivity. ıst is, in addition, influenced by risk. The comments below will ıerefore address the issues of risk and productity and possibilities of ıw to influence them.

ISK

Tunnels are affected by geology both as a foundation material and as a ɔad on the structure. Uncertainties in geology, which also become more ignificant with distance from the assessible surface, will therefore ffect tunneling more than any other construction activity. Uncertainties n labor quality and availability, in environmental constraints like imited working hours and access, and in institutional conditions (such as ontractual setups or bidding processes) will increase the overall ncertainty.

These uncertainties introduce risk. The parties involved in tunnel onstruction, namely, owner, designer, contractor (and labor), are all ffected by this risk. One can distinguish controllable risk, i.e., risk hich can to some extent be influenced by the affected party, and ncontrollable risk. On the other hand, risk can be handled by risk eduction and risk sharing. One will therefore attempt to reduce some of he uncontrollable risk and to share the remainder between the parties. his should be done in such a manner that the party best able to afford it hould carry most of the uncontrollable risk. A good example is risk aused by geological uncertainty which is largely uncontrollable. It can e reduced by exploration. The remainder should be carried by the owner s the party best able to afford it, but not exclusively to provide the ood contractors with possibilities to fully apply their capabilities. It s obvious that controllable risk, as for instance, risk associated with quipment performance uncertainty, should be fully borne by the party ontrolling it, namely the contractor.

RODUCTIVITY IN TUNNELING

A number of studies on productivity in U.S. construction (Paulson et l., 1977, Business Round Table, 1983) came to the conclusion that onstruction productivity has been decreasing for the past 10 to 20

ınneling and Underground Transport: Future Developments in Technology, Economics, and Policy
P. Davidson, Editor

years. A review of output per constant dollar, be that output in the form of linear foot of tunnel or tunnel volume, has shown a similar trend. In addition, advance rates have, after an increase in the early 1960s, not seen much change since then. The latter fact is particularly surprising given that available technology in form of tunnel boring machines and hydraulic drills, for instance, has made significant progress over the same time period.

While technogical improvements do have a favorable effect on productivity they cannot significantly influence productivity if introduced in isolation. The construction process (operation), constructibility (integration of design and construction), and institutional conditions need to be changed in conjunction with technological innovation. The above-mentioned tunnel boring machines and hydraulic drills provide a good example. By themselves they create the potential for a substantial increase in advance rates. The remainder of the construction process has, so far, not seen comparable improvements. This, together with the higher capital cost of the equipment, has often led to a smaller number of pieces of new equipment per project; the overall advance rate has thus not increased as much as some of the technological improvements would make possible (clearly, institutional and other factors like labor-employer relations and environmental restrictions play a role, also). A significant improvement in productivity can thus only be expected from radical changes in all factors affecting productivity. A final observation in this respect, which also brings the importance of a short construction time *per se* into focus again is the fact that many tunnels built fifty to one hundred years ago had similar total construction times as today's tunnels. This was achieved by using a much larger labor force than today and by having a construction process allowing this labor force to advance the tunnel from many points of attack.

It is left to the reader to take this last example, rethink tunneling and arrive at major breakthroughs in productivity increase and risk reduction.

REFERENCES

1. Ashley, D. B., Veneziano, D., Einstein, H.H., and Chan, M.H., 1981, Geological Prediction and Updatinq in Tunneling--A Probabilistic Approach, *Proc. 22nd U.S. Symp. on Rock Mechanics,* pp. 391-396.

2. The Business Roundtable, 1983, More Construction for the Money, *Summary Report,* The Business Roundtable, New York.

3. Einstein, H. H. and Vick, S.G., 1974, Geologic Model for a Tunnel Cost Model, *Proceedings*, Rapid Excavation and Tunneling Conference, Vol. 2, pp. 1701-1720.

4. Ioannou, P. S., 1984, The Economic Value of Exploration as a Risk Reduction Strategy in Underground Construction, Ph.D. Thesis, Massachusetts Institute of Technologv, Cambridge, Massachusetts, 461 pages.

5. Paulson, B. C. *et al.*, 1977, Development of Research in the Construction of Transportation Facilities, *Technical Report No. 223,* The Construction Institute, Stanford University.

nancing Macroprojects

STER C. THUROW
rdon Y. Billard Professor of Economics
oan School of Management
ssachusetts Institute of Technology
mbridge, MA 02139

Economics in many ways stands as the enemy of tunneling and macro-
gineering. The problems stem from the standard rate of return on
vestment (ROI) and net present value (NPV) calculations. Basically
ere is an interaction between interest rates and the length of time that
takes to complete a project that biases decisions away from undertaking
ngthy long-lived projects.

This phenomenon can be seen in American investment in plant and
uipment in 1984. Record high, inflation corrected, real interest rates
d not stop investment in private plant and equipment. Total investment
s near a record high, running at 11.6% of the gross national product
NP). But it was all short-term one or two year investment. Investment
long-lived facilities, plant or equipment was at a record low.
sically no new green field major industrial facilities were being built
America. High interest rates did not stop investment but they did stop
ng-lived investment.

The reasons for this are simple and can be seen in a simple table of
scounted net present values. As table 1 shows, with a 5% interest rate,
project scheduled to come on line in 5 years must expect net present
lue benefits of $1.28 in the fifth year per dollar invested in the first
ar of the project if the project is to be profitable. If the project
mes on line in 15 years the net present value benefits that must be
pected rise by two-thirds to $2.08. But at a 15% interest rate a 5 year
oject must earn $2.01 in benefits and a 15 year project must earn $8.14
benefits. The required expected benefits rise by a factor of more than
ur.

Looking at the net present value benefit calculations with a 5%
terest rate the value of a dollar's worth of benefits 30 years from now
$0.23, at a 10% interest rate they are worth $0.06 and at a 15%
terest rate $0.02. A threefold increase in interest rates essentially
pes out the current value of the distant future.

BLE 1. Benefits Necessary to Cover One Dollar of Initial Year Cost

	Length of Project		
terest Rate	5 years	10 years	15 years
Percent	$1.28	$1.63	$2.08
0 Percent	$1.61	$2.59	$4.18
5 Percent	$2.01	$4.05	$8.14

neling and Underground Transport: Future Developments in Technology, Economics, and Policy
. Davidson, Editor

Lengthy projects get eaten alive by interest rates. Thus, when you read that a nuclear power project costs $8 billion you are not really reading what the average person thinks they are reading. Typically, the fine print is read that $8 billion nuclear power project is composed of $2 billion in construction costs and $6 billion in interest charges accumulated during the period of construction. What is killing the nuclear power industry is not so much high interest rates by the time necessary to complete a project. If the electrical power generation industry had not been a regulated industry where the regulator passes interest charges on to the consumer, rising interest charges would have killed nuclear power long before they did.

Interest rates are not controllable by the tunneling or macro-engineering community, but time is. Cutting the length of time it takes to complete a project from 15 to 10 years is equivalent to a very large reduction in interest rates. At a 15% interest rate total costs are more than halved.

As a result, if the three words of wisdom in the real estate industry are "location, location, location" the three words of wisdom in the tunneling industry ought to be "speed, speed, speed." Whenever possible tunneling technology ought to move toward techniques which permit faster completion times.

From this perspective, data on completion times are very discouraging since they systematically show that it takes longer and longer periods of time to complete major construction projects. This, of course, means that fewer and fewer major construction projects are economically viable.

There are several reasons for this slowdown. In public projects there is a focus on minimizing construction costs, but this often raises total costs. Firms win contracts by having the lowest cost bid but this bid price does not have to include the accumulation of interest during the process of construction--these interest charges are directly paid by the buyer and are not part of the bills submitted by the contractor. As a result, if a dollar of construction costs can be saved by completing the project more slowly, the contractor has an incentive to do so even though the extra time may end up increasing interest charges by more than the construction costs saved. If contractors had to pay accumulated interest charges, they would operate very different construction projects in terms of the time necessary for completion. They would be willing to incur extra construction costs to speed up the project because what they spent on construction would be more than paid for by lower interest charges.

In tunneling, multiple faces may be inefficient in terms of construction costs but highly efficient in terms of total costs. Large crews may not be as efficient as small crews in terms of construction costs, but worth it if total costs, including interest charges, are included.

Learning curves may have disappeared in major construction projects. Certainly this is true in the data on the construction of electrical generating plants. In the 1950s and 1960s the United States built most assembly line plants with the same construction crew moving around the country to build plants. The result was a steep learning curve and good gains in productivity. In the 1970s the United States built one of a kind nuclear power projects where there were essentially no learning curves.

Shifts should be made to better tunneling technologies, but engineers need to think as to how learning curves can be obtained from tunnel to tunnel. This may require changes in work rules, changes in how work crews are recruited, or changes in bidding in terms of awarding contracts for more than one tunnel at a time.

In the long run, interest rates can only be reduced by raising savings rates. Given U.S. savings rates well below those of the rest of the industrial world one would expect real U.S. interest rates to be well above those of the rest of the world, as they are. As a result, whatever macro-engineering is done, is apt to be done outside of the United States.

In addition to taking longer, productivity has been falling (it fell at a 2.2% annual rate from 1977 to 1983) in the U.S. construction industry and now is about 25% below the peak levels of the late 1960s. In the 20 years before that construction productivity was growing rapidly. The reasons for this decline are a bit mysterious, but they spell disaster for tunneling projects. Some of the problem relates to the increasing number of permissions that must be obtained to engage in construction. But more design and engineer hours of work now seem to be required per unit of output. Projects also seem to bog down near their completion as workers slow down knowing that they are likely to become unemployed when the particular project that they are now working on comes to an end.

If one looks at the history of macro-engineering it is clear that most macro-engineering projects did not have to meet ROI or NPV calculations. They were undertaken as an act of faith or fear. Cathedrals and pyramids were enormous macro-engineering projects extending over many years. But they were not viewed as economic investments. They were a gift to God, a way to buy immortality, a passport to heaven--i.e., they were consumption.

Modern macro-engineering projects, such as the man-on-the-moon projects, were similarly done for reasons of faith and fear. Americans feared that the Russians would get there first and that there might be significant military advantages from being able to get to the moon. But they also went to the moon as an activity of faith and discovery. It was exciting. The man-on-the-moon project neither after nor before was asked to meet an ROI hurdle rate. At no point could it have done so.

Interestingly, modern man treats investment and consumption in radically different ways when analysing when it should be done. Investment should not be done unless it meets the tests of the investment calculus. Consumption needs meet no test other than that someone wants to do it. Thus, the man buying a recreational vehicle cannot justify its purchase based on the number of motel bills he will save, but he buys it anyway and no one calls him to account because he is not living up to the rules of hard nosed consumption analysis. Yet, one can easily argue that a tunnel not meeting ROI calculations is more valuable to society than a recreation vehicle that is not required to meet ROI criteria. To the modern man, investment is automatically treated as a sacrifice that must be repaid with future benefits while consumption is treated as a direct benefit that need not prove its worth in future benefits.

I suppose one question this leads to is how can tunneling be viewed as a consumption good rather than as an investment good. Under what circumstances is it an end in and of itself?

If one looks at existing macro-engineering projects and asks under what circumstances they paid off, it is clear that linkages and externalities become central. The Erie Canal never paid for itself. The railroads were built too soon after its completion and took its markets away from it before it had time to repay investors--a principle one being the State of New York. Yet one can argue that the Erie Canal was a good investment for the State of New York since it gave New York City a transportation cost advantage for a few years and those few years let New York get a lead on either Boston or Philadelphia as the country's industrial and financial hub. If the State of New York had not invested in the Erie Canal and waited for the railroads, Philadelphia might very well have had a transportation advantage that would have led it to become the nation's number one city. But by the time the railroad actually came, New York had already established such a lead that Philadelphia's rail cost advantages were not enough to allow it to displace New York economically.

If one looks at the western railroads it is clear that they would not have been good private financial investments without the heavy subsidies coming from government. The Northern Pacific Railroad, for example, got every other square mile of land in a 100-mile swath on either side of its rail line. It was profitable with that subsidy, but would not have been profitable without it. Yet the railroads west of the Missouri were a good investment if one remembers that they were what permitted farming. Without them the farm projects of the west could not be sold since they could not be brought to market. Here again a macro-engineering project paid for itself by its external effects on the economy of a region.

Consider the Rural Electrification Administration (REA). It needed and received subsidized interest to bring electricity to the farm since private electrical utilities could not profitably do so a market rates of interest. Viewed narrowly at market rates of interest the REA was a financial failure. Yet was it? Americans tend to forget that United States agriculture was not always dominant in world trade and did not always have a good rate of productivity growth. In the 19th and early 20th century, the world's principal grain exporter was Russia and not the United States. In the late 1920s, agricultural productivity was low and growing very slowly.

The REA and other farm programs essentially turned this around. With electricity, it was possible to electrify the farm and mechanize it in ways that would have been impossible without electricity. The result was 5 decades of rapidly rising productivity and the transfiguration of a sick industry into a healthy industry that became the world's dominant international exporter.

The REA paid for itself many times when viewed broadly, but never paid for itself if viewed narrowly. The payoffs came in linkages and improvements in productivity that could not be captured by the REA itself, but were captured by the American economy more generally.

The moral of the story, of course, is that in looking for profitable macro-engineering projects one should look for projects with similar linkages and externalities to the rest of the economy.

Tunnels and International Macroventures: Experience and Prospects

SIR ROBERT G.A. JACKSON, KCVO, CMG, OBE, AC
Senior Advisor to the Secretary General
United Nations
New York, NY 10017

I am very grateful to all those who asked me to participate in this conference. I am concerned, however, that you have asked the wrong man to talk to you and for several weeks I have been reflecting about this difficulty. In doing so, I recalled an incident with Mr. Churchill during the Second World War.

The Prime Minister had become very impressed with certain dispatches from the British Embassy in Washington and had inquired who was responsible. He was informed that they had been prepared by Isaiah Berlin --who will be known to many of you--and gave instructions that he was to be brought to Number 10 (the address on Downing Street which is the official residence of British Prime Ministers) immediately when he next visited England.

In due course, Berlin landed in an airfield in England and, to his surprise, was taken immediately by a security officer to the Prime Minister who, for the next two hours, talked majestically about the course of the war, his grand strategy, and the political problems that would arise after victory had been achieved. During the meeting, Berlin said little, except to agree at appropriate times and, occasionally, to make a generalized comment. After his departure, the Prime Minister remarked to his Principal Private Secretary, "Extraordinary! Here we have a man who sends us quite remarkable dispatches and yet during the last couple of hours he has scarcely made a comment to me."

Meanwhile, Irving Berlin, not Isaiah, equally puzzled, proceeded to Claridge's and met the agents who were responsible for arranging his various concerts in England!

Bearing that incident in mind, I hope you will be tolerant with me on this occasion!

I have been asked to talk of my experience with large-scale projects--particularly any that have involved tunneling--and in doing this I should like to place special emphasis on the critical phase of decision making. I will talk briefly about five projects with which I have been concerned personally, and then make a few general comments. If a title should be required for this talk, I would suggest a robust Australian expression "Let's give the bastard a go!"

First, let us look at the defense of Malta during the Second World War. In 1937, the British Cabinet's Committee of Imperial Defense--composed of very senior ministers and outstanding representatives of the armed services--ruled that Malta could not be defended. About the time of Munich, in October 1938, we received in Malta Robert Watson-Watt's second "Radio Location" set: the prototype of today's radar. As soon as we saw specks on the screen which were, in fact, aircraft operating from Catania and Syracuse, I was convinced that Malta could be defended successfully, for the R.A.F. would now have time for their fighters to gain operational altitude.

Tunneling and Underground Transport: Future Developments in Technology, Economics, and Policy
F.P. Davidson, Editor

We therefore decided to advance proposals to the top brass and for this purpose consulted with our colleagues in the other services and the government of Malta. Our case rested basically on the fact that Malta was made of solid rock and that everything could go underground, sustenance of course, depending on the Mediterranean fleet based both in Alexandria and Gibraltar, and air cover depending on the R.A.F. having sufficient time for the fighters to reach operational altitude, hence the decisive importance of radar. The Army, of course, was another essential element in terms of local defences and, as it turned out, also in terms of anti-aircraft defense.

With the cooperation of my admiral (my chief use to him was as a tennis partner!), a submission was made to the Commander-in-Chief, Mediterranean, who immediately proceeded to London in person. Within a week, we received a remarkable cable--which remains one of the high points of my life--which read roughly "All proposals for the defense of Malta approved. Parliament has not approved any funds for this expenditure but proceed with all possible despatch and where necessary quote MED 107." That decision decided the future of Malta, and many strategists would argue that the defense of Malta was at least a critical--and may well, indeed, have been the critical--factor in preventing the junction of German and Japanese forces, thus placing the outcome of the Second World War in jeopardy.

As to the tunneling related to that defense, it will be recalled that Malta sits on the floor of the Mediterranean, very much like a large mushroom, with a source of pure water at its base. The island is made of sandstone, which is easily worked, and one of our first jobs was to excavate seven oil fuel tanks--each 340 feet in length, 90 feet in height, with a 30-foot span, thus enabling us to store about nine months' consumption underground. Simultaneously, a power station was also built underground, with a strategic ring main extending all over the island. Thus, for at least nine months we had an invulnerable source of energy. The Romans had left us storage for grain in the shape of fosses, cut out of the sandstone, shaped like wine bottles, and each about 50 feet in depth. Here again, we were able to store nine months' supply of cereals in safety, and also placed a flour mill underground at the same time that we built the oil fuel tanks and the power station. Thus, with pumps also underground, we were certain of our basic needs of bread and water for nine months.

Simultaneously, we began to build underground shelters for the civil population in the most vulnerable areas--about 100,000 (plus a total population of about 340,000 people). Initially, they could just stand in the tunnels and then, by degrees, they would be able to sit down, then to lie down. In the later stages of the siege, babies were born underground and elderly people died there. Our communications systems, our hospitals, many of our workshops, and our submarines also went underground or had very heavy concrete shelters provided for them. In this vast operation only the R.A.F. declined to put their aircraft underground and, tragically, they paid for it.

In the event, the defense was successful, withstanding the most concentrated bombing of any target in the world during the Second World War. This was a great credit to the armed services, and the U.S. Navy also played an important and invaluable role in providing flat tops which enabled reinforcements of fighters to be sent to the Fortress and, above all, the Maltese--true to their magnificent tradition in resisting the siege of the Turks long years before--were a decisive element in the defense.

As to tunneling, we kept no records, but I remember that provision for the civil population alone required 23 miles of excavation.

I have talked at greater length about Malta than any of the other examples, to which I shall refer shortly, for it contains many lessons. First of all, the willingness to challenge accepted views, especially when negative and existing at the highest levels of government, required courage. Second, in order to challenge those views, imagination and vision were also required. Third, an element of luck--for example, the fact that radio location became available at a highly sensitive political time, i.e., the Munich crisis, was often vital. Finally, of course, is the willingness of the decision makers to grasp the nettles.

Let me now turn to the Snowy Mountains Project in Australia, which remains one of the joys of my life. As many of you know, Australia is a relatively flat and dry continent. For centuries rains have come across the Pacific from the northeast, descended on the Southern Alps in the southeast of the continent and then drained back into the ocean. It is nearly 100 years ago since the concept of reversing the flow of the Snowy River, which originally emptied into the Pacific, was advanced. By the end of the Second World War, the Commonwealth Government saw the project, which straddles the border of New South Wales and Victoria, as a great source of energy and of irrigation--the latter fitting in very well with the government's policy for immigration. For over 50 years the pros and cons of this great project had been argued, and by the time about which I am talking--just on 40 years ago--the cost was esimated at over one billion pounds, or about $10 billion in today's money. That may seem a small sum today, but at that time to Australia it represented an almost fantastic investment.

At last the project came to Cabinet under the leadership of a most remarkable Prime Minister for whom I retain the highest respect, Mr. Ben Chifley. By profession he was a locomotive driver and as Prime Minister remains one of the few genuinely great heads of government I have met during the last 50 years or so.

If I may take up your time here for a minute or two, I should like to comment that at the same time about which I am talking--shortly after the Second World War--I was Chief Executive of the United Nations Relief and Rehabilitation Administration--the first of the modern UN organizations. Among other things, we had 8,500,000 displaced persons on our hands in Europe. No government--the US, the UK, Brazil, France, any government you could name--was willing to accept any of them. Yet, when I put this enormous problem before Chifley, a year or two before the Snowy decision in Canberra, he immediately agreed to take at least 100,000 people (ultimately nearly one million people went to Australia) and thus broke the logjam, encouraging other governments in North and South America and other parts of the world to accept these unfortunate victims of war.

To get back to the Snowy story, the Prime Minister had the Commonwealth Director of Engineering--a most distinguished man--Dr. Louis Loder, on his right, and asked him, "Dr. Loder, will this bloody thing work?" "Yes, sir," replied Loder. "Then," said the Prime Minister, "let's give the bastard a go!" And that is how the Snowy came to life--or nearly so! That was a magnificent and courageous decision--one typical of President Truman--and is one of the very rare examples where one man demonstrated the all-important leadership and decisiveness that brought an entire Cabinet into line. Incidentally, General Smuts was another of that very rare breed of great leaders.

To be strictly accurate, I should add that Mr. Chifley's government shortly afterwards lost power and its successor, led by Mr. Robert Menzies, initially showed marked opposition to the project. In a relatively short time, however, the government was persuaded to initiate the project and the rest of the story is simply one of success.

In passing, I should add that at that time Arthur Morgan of the TVA had come into my life, and had impressed on me the very large dividends that could be derived from these great multi-purpose projects if tourism--in the widest and best sense of the word--was thought out carefully. Today, revenue from tourism in the Snowy region has an importance equivalent to the value of its power and irrigation.

The Snowy was expected to take 32 years to complete. Experience was gained, of course, as each stage was completed, and in the end the entire job was finished in 25 years, with an energy and irrigation capacity roughly double that which had been expected originally. Some 98 miles of tunneling were involved and in that process several new techniques which have now been adopted in other parts of the world were evolved. In the execution of the project, world records for tunneling were broken regularly by the work force made up of men from 44 countries.

Finally, I cannot emphasize too strongly the critical importance of Mr. Chifley's decision, and I have quoted it all over the world again and again in efforts to persuade politicians to act with courage and vision.

Let me now turn to the third example, the Volta Project in what was then the Gold Coast, now Ghana. From the point of view of the United Kingdom (with strong North American support), the primary objective was to secure a strategically secure source of aluminum production. However, from the point of view of the Ghanaians, the objective was quite different. The Gold Coast represented the very first phase of the British withdrawal from Africa: the Prime Minister, Dr. Nkrumah, saw the project as a great example of what an independent African state could do, as well as providing a second leg to the national economy which at that time depended almost entirely on cocoa production.

The project, from a political point of view, was exceptionally difficult, in that the British had recently undergone a disastrous experience with the groundnuts project in Tanganyika, and there was marked resistance in London and elsewhere to investing something on the order of $500 million in a highly volatile political part of the world. An even more difficult problem arose from Dr. Nkrumah's efforts to pursue a foreign policy which he regarded as genuinely nonaligned. Here I will refer only to the basic decision to go ahead with the project which today continues to produce the cheapest power in Africa.

In the ultimate analysis, the basic decision was taken between President Kennedy and Mr. MacMillan (now the Earl of Stockton), and in that process my wife's close friendship for over twenty years with the President on the one hand and my own friendship with Mr. MacMillan on the other, were very real factors. So, in the case of the Volta, the luck factor came in again, in the form of personal friendships. Edgar Kaiser should also be mentioned here. He and three other contractors from the West Coast won the second contract for the development of the Snowy, and our joint involvement at that time led on to his playing a major role in the Volta Project.

For those of you interested in a detailed record of how this project came to life, I recommend a very interesting book _J.F.K.: Ordeal in Africa_ by Professor Richard Mahoney. By coincidence it was published the day after Frank Davidson's excellent book Macro.

At the end of it all, President Kennedy wrote to Barbara, my wife, commenting "We have placed quite a few chips on a very dark horse". The dark horse, sadly, disappeared from the scene, but throughout all Ghana's political turmoil in the last 20 years, the Volta River Authority has gone ahead successfully and has paid interest and capital on every loan on the nail.

As to my fourth example, the Channel Tunnel between England and France, I shall make only passing reference. This is the perfect example of not giving the bastard a go. As Frank Davidson has commented in his book, Napoleon and John Fox were in agreement about this project 183 years ago. Now we observe M. Mitterand and Mrs. Thatcher making another attempt to bring this great scheme to life. At the moment--as has happened so often in the past--prospects are a little brighter, but it is heartbreaking to observe the continuous lack of political decision and the subordination of vision and opportunity to a passing political philosophy.

My own interest in the Channel Tunnel has gone on for about 40 years, primarily as a result of my close friendship with Leo D'Erlanger who for many years was Chairman of the group seeking to activate the project. Knowing that his great-uncle had held a similar position in 1894, Leo frequently reflected on the number of generations of his family that would require to be bred before a decision could be taken!

One can only hope that this enterprise--so obviously right--will soon come to life. If the world ever needed a shot in the arm in terms of constructive and imaginary enterprises--as opposed to the monotonous and depressing development of weapons of destruction which, if God is kind to us, will never be used--this is it.

Finally, there is the Sahara. The recent tragedy needs no description. In October 1949, representatives of various Metropolitan and African governments met in Pretoria and created an Organization for Scientific Cooperation south of the Sahara. Subsequently, talented men and women worked in that organization, where the first Secretary-General was Barton Worthington, who had been my scientific adviser in the Middle East during the war, succeeded by Paul Marc Henry and Claude Cheysson. From the beginning many people tried to interest the world at large in the relentless movement of the Sahara to the south. To me it has always represented, as it were, a glacier of sand of incredible magnitude remorselessly moving forward, and rendering futile any efforts made by men to resist its progress.

Dr. Davidson has dedicated his book to "The Romans who demonstrated that the Sahara Desert is not invincible." This was undoubtedly true for the Romans, but for my generation there is overwhelming evidence that the Sahara is winning the present battle. My judgement of current efforts to deal with the problem is that they have virtually no chance of success. And so the tragedy may well be compounded.

To sum up, it seems to me that the first requirement for these great undertakings is vision and imagination. These are rare qualities but they do exist and the current scientific and technological revolution is constantly opening new horizons.

The second requirement is for the people affected by the project to be convinced that it is in their own interests for it to be brought to life.

It is here that political, economic and environmental education, as well as courage and persistence, are decisive factors.

Then comes the most critical phase of all--the political decision. Here the record is far from encouraging, as we all know, but I take a grain of comfort from the belief that the scientific and technical revolution will increasingly force political leaders to make the right decisions.

Today we live in a world where the dominant force is fear, leading to insane and negative expenditure on weapons of destruction which, as I have said, one can only pray will never be used.

The challenge today to all of us--no matter what our station in life--is to reverse that negative trend and to arouse the best and most creative instincts in our fellow men, never to give up, and to pray that more Chifleys will appear, prepared to "give the bastard a go." Now more than ever, we need that vital element of luck.

Introduction to Mock Hearing by Session Chairman

PETER E. GLASER
Vice President
Arthur D. Little, Inc.
Cambridge, Massachusetts 02139

Please take your seats so that we can proceed with our afternoon session. Sir Robert Jackson asked me to make one further comment regarding his speech. He wanted to make sure that you realize that the Mount Snowy Tunnel was 98 miles long, so that was a major accomplishment.

This afternoon you are going to look at the policy issues, the environmental and societal challenges. So far in this conference, we have heard some very interesting papers. Our focus was primarily on the technical. I would like to echo Lester Thurow's statements and admonition that the three things that we should focus on in terms of our completion of macro-engineering tunneling projects are speed, speed, and speed. I tried to listen when Lester Thurow was talking, and I heard two of those explanations for speed. One was technology. The second was productivity. But I did not hear the third. I believe Sir Robert alluded to it--that is, the speed of decision. This afternoon we will look at macro-engineering projects which we certainly believe are technically feasible. But it is the speed of decision making, sometimes under the control of the project manager and at times not, that we will address as our witnesses will be taking their places.

Traditionally, when we talk about a macro-engineering project, it is conceived and executed with a definite start date and agreed-upon performance objective or objectives, budgets, schedules, and there is an identifiable management organization which is held responsible for its implementation. It is rather easy to talk about success or failure since that is judged by whether or not the project's objectives are met, within budget and on schedule. Now, the time schedule comes to the speed issue. It is really not quite under the control of the management because the projects tend to be vulnerable to litigation. If changes in the regulatory environment occur, and if disputes extend over a decade or more (which is not that unusual with some of the projects we heard about), they are vulnerable to changing economic and political conditions. Furthermore, we know that any such major projects require a continuing consensus between public and private investors as well as the support of appropriate interest groups and government agencies at the various stages of the project. The emergence of what we now know as a familiar feature when we talk about large projects is the political culture that has led to a broad public involvement in policy issues related to economic, environmental and societal effects. As a result, some policy issues which required highly specialized input and which were at one time the province of the experts, the engineers, the technocrats and managers who could dispassionately in small numbers make the necessary decisions, are now the focus of citizens' interest groups of various kinds, of various ideological persuasions, and therefore often the subject of emotionally charged discussion. This phenomenon has in recent years affected public policy related to macro-engineering projects in part because of the negative view of technology that has been the hallmark of the 1970s and partly because now we know more. We have increased our understanding of undesirable economic and environmental effects on society. We have lost some of our innocence and our advocacy and at times we have been surprised by results which we could not plan for and which were quite unexpected. Today it is indeed a daring manager of a macro-engineering project who would neglect such issues, even

Published 1987 by Elsevier Science Publishing Co., Inc.
Tunneling and Underground Transport: Future Developments in Technology, Economics, and Policy
F.P. Davidson, Editor

at its earliest stages. At the conceptual stage, we must start to include an examination and understanding of the technical, economic, societal and environmental issues, and public attitudes when we are starting a project. We can clarify these issues during the planning stage, for if we leave that to the later stages, we may find that they become known by such names as Seabrook. It is obvious that we have to tradeoff environmental risks and costs against the perceived benefits. Whenever we talk about a macro-engineering project, the decision not to proceed with the project could also have undesirable impacts, whether they are ecological or economic. For if a project is not developed, some other route which is clearly more economically attractive, environmentally designed, and socially desirable must be found. It must be accepted that a risk-free application, particularly of new technologies, is nearly impossible. The best that can be done is to choose a course which maximizes benefits while minimizing undesirable side effects. Environmental impacts must therefore be considered in the light of alternative technologies which might be used to meet a project objective. Several planned tunneling projects will be of a scale that compares with the very largest engineering programs ever undertaken. The implications for society are profound. But whether or not such effects will be beneficial will depend upon political, organizational, and institutional factors. Proponents may believe that a given project would stimulate economic growth and that indeed a major contribution could be made to society. Opponents may hold that the same project, particularly one which affects many people, is a pernicious example of megatechnics. We must be wary of preoccupation with technical feats, which the public at times holds responsible for the unsatisfactory predicament of society. It has even been claimed that to accomplish the objectives of some of our macro-engineering projects would require creation of an authoritative state to protect massive capital investments and to force people to accept exploitation by various corporations. It seems clear that the objective of a just and rewarding society is more readily obtainable by maximizing the options and opportunities available to all people than by curtailing them or coercing them into adopting a particular and perhaps less desirable lifestyle.

This afternoon, rather than hold forth in various prepared papers, we have arranged what we call a mock hearing. The mock hearing is not going to be quite so "moot" because the participants are rather familiar with such hearings. I am delighted that Senator Paul Tsongas agreed to chair this afternoon's hearing and to have some eminent witnesses appear before him.

Introductory Remarks to Mock Hearing

PAUL E. TSONGAS
Foley, Hoag & Eliot
Boston, Massachusetts 02109

Let me welcome you to Massachusetts. I sat in on the last part of Sir Robert Jackson's presentation, and I found it fascinating, because I believe that in any kind of economic development, decision making is the most critical component. I come from the city of Lowell--some of you know of it and some of you will not. The city has made the front page of the New York Times, the Washington Post, and the Wall Street Journal as the success story in terms of middle-sized city "urban renaissance." And the reason it has happened there rather than someplace else is very simple. Decision making, the process of decision making, has been elevated to an art form: if you come into the city and want a decision on a whole host of issues, you will get it before you leave. That, I think, is the single biggest reason Lowell is where it is.

In addition to being a lawyer I am also a developer these days. My development partnership is in industrial parks. When we decided where we wanted to go, there were two issues for us. First came the economic issues, but equally important has been the judgement, mainly mine, as to which communities have decision-making capacity. Where we detect that capacity and a willingness to move, we make our efforts. Where we think that capacity does not exist, we stay away. And I suspect you are in the same situation.

The final example I will tell you about, and those of you who are from here will understand this. Here in this city we have the oldest functioning sports arena in history. It is called Boston Garden, where the Celtics, the Bruins, and a number of rats have their home. And I have been involved in an attempt to build a new arena for the last four and a half years. If you go down there, you will find that nothing has changed except the rats have matured considerably during that period of time. It is no accident when you compare what has happened here and what has happened elsewhere that during that exact same period of time, neither government which we have had has taken a public position on the issue. That is not mere coincidence. Like every other Senate hearing, this one is characterized by the fact that the Chairman knows less about the issue than the panelists, and we will continue that tradition here as well.

The witnesses will introduce themselves, but let me just say Governor Hickel and I know each other somewhat. I was on the opposite side and introduced the competitive legislation, the Alaska Lands Bill, which passed, and I have also been reasonably involved as an attorney with the Humane Society on some seals in the Pribiloff Islands where again we're on opposite sides.

Published 1987 by Elsevier Science Publishing Co., Inc.
Tunneling and Underground Transport: Future Developments in Technology, Economics, and Policy
F.P. Davidson, Editor

Planning Macroprojects

WALTER J. HICKEL
Chairman
Yukon Pacific Corporation
Anchorage, Alaska 99510

Alaskans must be forgiven if they cite the frontier--and the concept of the frontier--as part of our present reality. The Far West may have been reached and settled more than a century ago, but even today the Far North needs further exploration, comprehension, and settlement. When I served as Alaska's Governor, and later as the United States Secretary of the Interior, I developed a keen appreciation of the need for both information and determination: information, because inappropriate and untimely action can damage the very values we wish to preserve; but also determination, because recalcitrant problems require not only research, but action, and action can only be effective if it is based on decisions that command continuing allegiance.

I trust that the history of the Alaskan pipeline will prove both instructive and useful to the deliberations of this assembly. The project was the subject of bitter controversy for many years--and, indeed, the debate about environmental impacts may have improved some aspects of engineering design. But what galvanized the decision to go ahead was the oil crisis, brought about by the OPEC cartel when skyrocketing fuel prices threatened the well-being of the American economy. Once a decision was made, for reasons that were crystal-clear to the public, work proceeded, and a vital resource--the oil of Prudhoe Bay--was made available via a pipeline more than 800 miles long--for the markets of the "lower 48." Today, however, we are called upon to galvanize public opinion on behalf of major projects which respond to long-term needs, even in the absence of a crisis that can make those needs dramatic and highly visible. It is a task worthy of our statesmanship, of our prescience, and of our experience as pioneers in what was quite properly called "the New World."

Before this conference ends, we shall be privileged to witness a demonstration, on the M.I.T. athletic field, of supersonic flight through an evacuated tube. Many of us recall the short story, written about 1895, in which Jules Verne imagined a "subway" carrying passengers in comfort and safety from Liverpool to Boston at "1750 kilometers per hour." This very morning, Dr. Robert Salter gave us the up-to-date engineering and scientific rationale for such a service. I do not know whether or not this will be the optimal solution for long-distance travel and transport, but I am impressed by the array of solid technical knowledge and experience which makes such a system thinkable. In the continental United States, with airways and surface arteries increasingly congested, supersonic underground transport must surely count as an alternative to be given serious and sustained consideration. I trust that those who promote this inquiry will bear in mind that our vital traffic is not only east-west, but north-south as well.

Present statistics indicate that our economy is slipping relative to the impressive track record of our leading competitors. May I remind you, therefore, that our competition, in Japan and elsewhere, has learned how to listen to our wisest industrial and academic leaders and to profit from their advice and counsel. America has always thrived on competition: do we not owe our Apollo program in no small part to the emergence of "Sputnik?" And cannot we coalesce our energies and our determination around a series of major undertakings that will utilize not only "high

Tunneling and Underground Transport: Future Developments in Technology, Economics, and Policy
F.P. Davidson, Editor

tech" but also that oft-neglected substratum of "know-how" in our vital traditional industries and crafts?

The late Harold D. Lasswell was fond of urging his classes to establish "decision seminars," i.e., meetings where responsible leaders would come together in order to acquire information about specific problems and opportunities and then to make and implement decisions. In this spirit, the indomitable Sir Robert G.A. Jackson, the naval defender of Malta in World War II and now the senior advisor to the Secretary-General of the United Nations, presided over a meeting of engineers a few months ago in order to develop a practical program for bringing a massive and reliable water supply to the drought-ridden Sahel. Practical people, willing to invest their time and devotion, can use the findings of research to advance the human condition: it is our attitude that must be regarded as the decisive factor; research on social and environmental problems must lead to solutions and activities, not endlessly to additional academic treatises and conferences, valuable as they may be at times. I welcome the emergence of interprofessional groups such as the American Society for Macro-Engineering, so that we shall have a forum specialized in the identification and selection of the great enterprises capable of galvanizing the American future and the future of our increasingly interdependent world.

This conference has provided us with a broad view of tunneling technology as it is moving ahead today. I have noted with interest that the civil and mechanical engineers have now been joined by nuclear and aeronautical engineers and also by computer scientists in pushing the "state of the art" closer to the day when tunneling can be a completely automated process. In the mining industry, there are many companies which routinely build more than 500 miles of tunnels each year. Concern for safety, as well as for economics, has taken the mining industry several steps forward on the road to automation. At some point, I have no doubt that tunneling will be economical enough to sustain the ultra-rapid transportation whose principles Robert Salter and Thomas Stockebrand have outlined to us with such conviction and cogency.

Essentially, a tunnel is a conduit, and as we leave this gathering, I am sure we will look upon that broader field of conduit technology as providing an appropriate cadre for further deliberations. Many of you know that my young colleague, Mead Treadwell, and I have been engaged in an effort to secure the many permissions needed to construct a gas conduit from Prudhoe Bay to southern Alaska. Building pipelines across rough terrain under adverse weather conditions is one of the solid achievements of modern technology. The concept of a worldwide distribution network for power and fuel is now as timely as what was, a generation ago, the novel concept of worldwide telecommunications. And just as we have accommodated the idea of communication satellites orbiting the earth, so shall we accept the logic of an eventual intercontinental grid for water supply and for bulk commodity distribution. Not everything will go via tunnel. But conduits we shall have, some underground, some on the surface of the earth, and some in the form of long pipelines laid on the seafloor or suspended below the surface of the sea.

In a recent disquisition on the future of space exploration and settlement, I stated that "the immediate challenge before us now as Americans is to allow ourselves frontiers again." Every new thrust of technology opens a new frontier. First of all, there is the frontier within our own minds. Once we can generate new and confident concepts, we can transform the outer environment. Macrotechnology will very soon give us the possibility of linking up the "global village" for planet-wide transport and supply. In this frontier task, Alaskans can perhaps play a pivotal role. We have concepts for the cooperative development of Arctic resources, including the idea of an eventual city built on the North Pole! In the more immediate future, the United States can improve its balance of trade by more than a billion dollars a year simply by agreeing to sell Alaskan gas to Japan: when obvious and sensible steps fail to be taken, our economic plight is bound to worsen. The idea of sending Prudhoe Bay gas to Chicago (President Carter's forlorn hope) is simply not economical: the natural markets for Alaskan oil and gas are in the Far East. Cannot our diplomacy see this situation in terms of opportunity, and arrange imports of Mexican petroleum products in a manner that will strengthen hemispheric solidarity? Where is the old Yankee shrewdness? Why do we allow the festering of this self-inflicted wound?

In a few months, Alaska will have the honor of hosting an international conference to launch a "Global Infrastructure Fund." Is it not consistent with this noble objective--and also very much in our own national interest --to think in terms of a North American infrastructure? I have referred to the need for north-south networks within our own continent. If one takes the adverse trade balance of Mexico and the United States (because, let us admit, we are both debtor nations), it would be to our mutual advantage to increase, at one stroke, the receipts we can garner from petroleum product sales to the Far East, while we relieve our neighbor, Mexico, of its penury and its distrust of the future.

Much has been said about oil. But the next century will surely see a proliferation of giant enterprises transferring water from regions of surplus to regions of deficit. I am mindful of the fact that the world's longest true tunnel transports water--80 miles from the Delaware Water Gap to the island of Manhattan. Aqueducts of various designs and dimensions will be needed to avoid impending water shortages in mid-Canada, in the midwestern and southwestern United States, and in northern Mexico. If the ancient Romans were able to turn North Africa from a desert to a garden--and in the teeth of a Sahara which was substantially the same as it is today--can advanced industrial nations like ours sit by and idly allow the early maps of the mid-continent to resume their primacy with the now discarded appellation, "Great American Desert?"

Americans have been a litigious society. But whenever a crisis has loomed, we have teamed up to meet it. Today's economic challenge can be met if we deploy our traditional skills in a series of astutely chosen major economic enterprises so that our infrastructure is truly "second to none."

In the back of the hall, I noticed a model of a "palleted automated transit" system (PAT) designed by Professor David G. Wilson of the Massachusetts Institute of Technology. By placing motor vehicles on pallets which are then moved on an automated guideway, eight to ten times as much traffic could be accommodated on a single lane of highway as at present. The system would save lives, reduce environmental pollution, and make it feasible to move cars and trucks in subway tunnels, thus improving both traffic flow and air quality on the surface. Is this not a possible answer to such recalcitrant problems as traffic in our crowded Northeast Corridor? Is this not an American technical system which could improve traffic conditions both within tunnels and on the surface? Here again, a man's mind has created a new technological frontier.

We in Alaska support the ongoing search for better methods of transport and communication. We have noticed that the main limits to growth lie within the human mind, and the world must therefore seek out and learn to design and prepare major cooperative endeavors that will improve the "built environment" as well as the larger biosphere on which all life depends. I hope that one upshot of this gathering will be a renewed resolve to build programs of education so that we may have a trained cadre of competent engineers specializing in the management of major projects. The vast conduits mentioned in previous chapters will require major investments of manpower, money, and materials; the impacts will be diverse and far-reaching; and the complexity of "managing" such affairs boggles the mind. Just as France created its system of "grandes écoles" to prepare for the industrializing wave of the 19th century, should we not move forward swiftly to build a graduate academy of engineer-management, worthy of the best accomplishments of American industry and devoted to the formation of a corps of able, experienced men and women who can evaluate, prepare, and provide the helmsmanship for those great enterprises on which the future of this nation, and of its partners on the planet, surely depends?

At least 100 new, independent nations were created in the aftermath of World War II. The vocabulary of international politics has been mired in a sterile rhetoric of "north versus south" or "east versus west." Inexorably, we are becoming an intertwined world. A graduate management college for engineering macroprojects, such as I have suggested above, could provide the intellectual and moral coherence that our fragmented world requires. And within each region, we need better institutions for cooperative investment. Even in North America, which has given so much good advice to the founders of the new European Common Market, is there not an opportunity, if only on an informal basis, for a North American consultative assembly that will help to harmonize our infrastructure and our policies?

I salute this conference for its pioneering effort to bring to interprofessional attention the new frontier constituted by underground space. Quite clearly, we shall soon have the means to dig tunnels more quickly, more economically, and with greater safety. And we shall have rapid means of transport to install the new underground networks. Let us hope that, after due debate and study, we shall have the cohesion and the courage to come up with those key decisions which can lead us, once more, on the path of prosperity and progress.

Regulatory Planning for Transportation Tunnels

COLONEL CARL B. SCIPLE
Commander and Division Engineer
New England Division of the U.S. Army Corps of Engineers
Waltham, Massachusetts 02154

There have been many impressive projects addressed at this conference, presenting both the latest in technology and analysis of the economics involved. In addition to the real "nuts and bolts" issues involved in macroprojects, designers must also address environnental considerations. The National Environmental Policy Act of 1969 (NEPA) served as the legislation requiring environmental reviews of the type of projects we have been discussing. Since the passage of NEPA, many levels of environmental control have emerged and increased in significance over time. With these controls have come an equally large number of regulators (local, state, and federal) who control or direct impacts on the environment according to a rule, often through the issuance or denial of a permit for the work. If I can only impart one clear message to you today it would be to not underestimate the work generated by the regulatory requirements. Substantial effort often is needed to get these permits.

To treat environmental controls lightly in project design is a mistake because they are important. Over the last two decades, the public has demonstrated a willingness to spend the funds needed to address environmental concerns. The more than a billion dollars spent on the initial "Superfund" activities needed to clean up hazardous wastes in this country is a good example. Regulatory factors (controls) are here to stay and securing the necessary regulatory approvals through permits is vital to a project.

Macroprojects must be able to pass a test on the effect they will have on the quality of life as well as their engineering sufficiency. Experience has shown that projects are often better after environmental reviews and subsequent design iterations. Consider the $800 million Fort McHenry Tunnel Tom Kuesel described, $300 million of which was the cost resulting from regulatory review, in other words, the cost of the permit.

Why is the consideration of regulatory planning in transportation tunnels important? Quite simply, to save the project planner time, effort and money. The Pittston Oil Refinery proposed for Eastport, Maine was in planning for over 10 years at a cost of $10 million and it was never built. A large shopping mall proposed for North Haven, Connecticut has been on the drawing board for over 10 years, with millions of dollars spent by the developer and there are still permits required before the first shovel of dirt can be turned. For a proposed mall in Attleboro, Massachusetts the developer has had to spend over $3 million dollars addressing environmental concerns and they are not out of the woods yet. Even a relatively small project, a half-million dollar diversion dike on the St. Croix River in rural Maine, took the developer over two years to obtain permits from more than 20 agencies. There are many other stories like these such as the development of Marco Island and the proposed Westway highway in New York City.

During this session, I will give a personal perspective of regulatory planning as I see it from my "regulator" role. The process can be a virtual minefield and I will try to show you where most of the large mines are. It's difficult to anticipate all of the regulatory considerations

Published 1987 by Elsevier Science Publishing Co., Inc.
Tunneling and Underground Transport: Future Developments in Technology, Economics, and Policy
F.P. Davidson, Editor

you may encounter but if you approach the process systematically, it will be a smoother road.

I will begin this primer on regulatory planning with the basics according to Mr. Webster. To regulate is to control or direct according to a rule. For the U.S. Army Corps of Engineers, the rules we apply are Section 10 of the Rivers and Harbors Act of 1899 and Section 404 of the Clean Water Act of 1977. We control activities under these rules through the requirement to obtain Corps permits. Of the 41,000 multidisciplined civilians employed by the Corps of Engineers, about 900 are directly involved with our permit review process. Obviously the staff you devote to your regulatory planning will be considerably smaller but there are some tips that will give you more mileage from those planners.

First, conduct research of regulatory requirements early on. There could be many and they must be anticipated. The December 21, 1984 edition of the Federal Register contains 32 pages that list areas of environmental concern and the agencies to contact about them. Use the agency regulators throughout the process. They can save you a lot of unnecessary effort.

Second, make a thorough assessment of environmental and other concerns related to the project. Measure or make a good estimate of the degrees of opposition lurking out there. Judge the validity of those concerns and how you can deal with them. Try to understand your potential opponents, and assess the local sensitivities that may arise.

Finally, consider the big picture and do not be too narrow in the scope of your assessment. Realize that ours is a public interest review. Even though our expertise, for example, is water and wetlands, the NEPA process mandates we consider all factors and make our decision based on the public interest. As a result, though we may be considering a permit application to place fill in the wetlands, we must consider the socioeconomic impacts of the proposed development as well as the direct effects of the fill to be placed.

After the project planner assesses the situation, one possible result might be to kill the project. For example, it took Con Edison over 10 years and a substantal financial investment before they reached this decision in the Storm King Pumped Storage project.

I believe we can agree that regulatory review does affect a project and can play a very pivotal role in the iterative design process. Positive changes often result from the reviews. Examples are, the preservation or creation of wetland; the protection of endangered species; the safeguarding of archaelogical sites; improved flood control and noise abatement. All of these lead to greater public acceptance.

Who in your organization should perform the regulatory planning function? A seasoned professional. Someone who has already "run the gauntlet." Ideally a multidisciplined team with one leader should be assigned to the task. Consider whether you want this work to be done in-house or by contract. If it is done out-house there should be a full-time staff person directing the effort by the contractor.

Whether or not you use your own staff, support the regulatory planning function--give it a priority and see that the effort begins early. What should your regulatory planning project team do? They should contact all regulatory agencies with possible interest in your project early. Find out who is interested, what their requirements are, and the procedures to be followed.

Request a preapplication meeting before applying for a permit. Have as many relevant agencies as possible at the meeting at the same time. This concurrent approach to regulatory planning saves a lot more time than if you approach the process sequentially.

Alaskan Governor Hickel, in discussing the Alaskan pipeline, noted that "environmentalists raised good issues." Have a positive approach and be ready to cooperate. Do not piecemeal the data. Package all actions that will need regulatory approvals and give them to the agencies up front.

Whenever possible, meet the decisionmaker--the person who will actually sign or deny the permit. Introduce yourself and the project. Elevate any questions or issues to the decision maker that you cannot resolve with the staff. Extend the project's critical path scheduling into preconstruction stages. Planners must know at what point each permit will have to be issued and what the life span of the permit is. Most permits expire after a certain time. Our permits, for example, generally provide three years to complete the project. Delays in getting new permits issued can have far-reaching impacts on interdependent elements of the project. Realize that though regulators want to process permits expeditiously, there are many potential hurdles. The Corps of Engineers has a goal of reaching decisions on permit applications in 60 days but delays are often caused by insufficient information being provided, public opposition, required inter-agency coordination, the preparation of Environmental Impact Statements and procedural matters.

Remember that during the Public Interest Review, unexpected concerns and opposition often appear. It is Murphy's law in action.

Let me conclude with a few final thoughts.

Environnental controls for tunneling projects and other project types are here to stay.

The regulatory process is participatory democracy at its best. People will get involved, not often in national policy issues, but always in local project development. The regulatory procedures provide an opportunity for varied interests to participate and they do!

Project planners should not procrastinate or skip meeting regulatory requirements. This most often only results in delays costing time, money and maybe the entire project. Meet the challenges of compliance head on. If you or the regulator skirt the process, you, the agency, or both will likely end up in court which, at best, could delay the project completion even further.

Coordinate early with the pertinent regulatory agency. Remember that they are not assigned to stop development, only to ensure compliance and that the public interest is protected. Treat the regulators as part of your planning team, working to get the job done.

Technology has made many exciting tunnel and other macro projects possible. There are many benefits to these advances but there is a price to pay in achieving those benefits. The price calls for coordination and sensitivity to others' concerns. The regulatory process is a helpful system to address those concerns through effective coordination and then design iterations. The changes often lead to better, more sensitive, and acceptable design. Most of all, the projects are responsive to the general public interest.

I firmly believe that both the project and the public interests are served by the regulatory process. It is an integral and complimentary part of the design process.

The regulatory role of the Corps of Engineers versus the Coast Guard for transportation projects in "navigable waters of the United States."

Each agency has separate regulatory authority under the Rivers and Harbors Act of 1899. The Corps would approve tunnels under Section 10 of the Act. The Coast Guard would approve bridges and causeways under Section 9 of the Act. Thus a project such as the Chesapeake Bay Bridge Tunnel between Virginia and Maryland would require permits from both the Corps and the Coast Guard. Further, the Corps would also have to permit the discharge of fill material used in the construction of causeways and bridge approach fills under the Clean Water Act and the Coast Guard would have to approve aids to navigation such as lighting and channel markers.

Question: What is the role of the Coast Guard versus the Corps of Engineers for transportation projects crossing "navigable waters of the United States"?

Answer: The Corps must approve tunnels crossing under the waterway under Section 10 of the Rivers and Harbors Act. The Coast Guard must approve bridges and causeways crossing over the waterway under Section 9 of the Rivers and Harbors Act. The Corps would also have to approve the discharge of dredged or fill material and the Coast Guard approve aids to navigation.

Role of the lead federal agency.

Many major transportation projects requiring permits from the Corps of Engineers are funded by the Federal Highway Administration of the U.S. Department of Transportation. The Council of Environmental Quality Guidelines have established a lead agency concept for the preparation of environmental documents when more than one federal agency is involved in a group of actions directly related to each other. In these instances, one federal agency, usually the one with the major involvement, will take the lead for preparation of an Environmental Impact Statement (EIS) with input from the other agencies.

Question: If Federal Highway Administration is funding the project and the Corps issuing a permit does each agency prepare an EIS?

Answer: No. One federal agency will be determined to be the lead agency. Most likely the Federal Highway Administration because of the funding and project type. They would prepare the EIS with input from the Corps as a cooperating agency. This would satisfy NEPA requirements for the environmental document for the permitting process.

Disposal of material excavated from a tunnel.

The Corps of Engineers regulates the disposal of dredged material in waters of the United States and their adjacent wetlands. The EPA is responsible for regulating the disposal of all other waste materials. Dredged material is defined as material that is excavated or dredged from waters of the United States. Since a tunnel could be excavated under a waterbody, essentially in the dry, it would not be considered dredged material. However, if it were to be disposed of in a water or wetland of the United States with its primary purpose being to create fast land on which to erect a structure, this would be considered the discharge of fill material requiring a permit under Section 404 of the Clean Water Act.

Question: Who regulates the disposal of material excavated from a tunnel?

Answer: Normally such material would be considered waste regulated by EPA. However, if it were to be disposed of in a waterbody or wetland with the intended purpose of creating additional land or an area on which to erect a structure it would be considererd the discharge of fill material. Such an action would require a permit from the Corps of Engineers under Section 404 of the Clean Water Act.

he Story of a Model Evacuated Tunnel for the emonstration of High-Speed Transport

HOMAS STOCKEBRAND
dvanced Engineering Center
igital Equipment Corporation
lbuquerque, NM 87122

The rather terse report on this page summarizes the actual low
riction high-speed transport model that was demonstrated at M.I.T. in
pril of 1985. Following this is the story behind the demonstration,
hich failed to work right up to the final shot with the people watching.

The demonstration was of a light gas gun with what may have been the
ongest receiver in the world: however it is certainly not a breakthrough
n any other way. That distinction must accrue to experimenters of the
uture.

OAL

The goal was to give an indication to the layman and professional
like of how necessary a vacuum is to reduce the energy requirements of
igh-speed transport.

ETHOD

The method used sought to show that a light projectile moving at
peeds above the speed of sound (in standard conditions) slows down far
ess in a vacuum than when it is projected into the air. The initial
elocity and difference in distance traveled was measured in the two cases.

ESCRIPTION

A 50-millimeter inside-diameter plastic tube forming a tunnel about
90 meters long was laid out in a straight line and evacuated to 1
illimeter of Hg (about 1/1000 of an atmosphere). A light gas gun
rojected into the tunnel. It consisted of a 40-millimeter inside
iameter plastic tube forming a barrel 1.7 meters long with breech 0.75
eter long of 50 millimeter tube. The breech was separated from the
arrel by a diaphragm of .075 millimeter cast acetate. The projectile was
ping-pong ball, 38-millimeter outside diameter filled with urethane
oam, just ahead of the diaphragm.

Helium was injected into the breech (mixing with any air already
resent) until the diaphragm broke at approximately 300 kPa (45 lb/in^2)
roducing a shock wave which sent the projectile on its way. The initial
elocity was measured by timing the time of transit between a pair of
hotocell sensors spaced a meter apart. The straightness of the pipe and
egree of vacuum were improved until the projectile in fact traversed the
ube.

The gun assembly was removed from the vacuum tunnel and the 3 gram
rojectile was shot into the ambient atmosphere.

ESULTS

The original velocity exceeded the response of the measuring
hotocells which had been shown to be 550 m/s. The upper limit of its
peed would be the speed of the shock wave which was calculated to be

Published 1987 by Elsevier Science Publishing Co., Inc.
Tunneling and Underground Transport: Future Developments in Technology, Economics, and Policy
F.P. Davidson, Editor

about 746 m/s. The degree of the vacuum was never known. The projectile shot through the air came to rest in 40 meters. (The speed of sound in standard air is 313 m/s.) It should be mentioned that in the real train, the accelerations of 10,000 "g" will not be acceptable!

THE STORY

The whole project was a long rambling string of near-misses ending with near failure, taking place on weekends from September 1984 through April 1985. The highlights of prior work are outlined below, followed by notes on the actual week of work in Massachusetts. The Appendix contains four reports made to the M.I.T. Macro-Engineering Research Group as the work progressed and form a sort of diary that may be interesting.

EARLY WORK

Programming the microcomputer to count microseconds between photocell interruptions took weeks and in the final shot it did not work. But it was invaluable in getting the apparatus running.

A great evening occurred when we first shot the gas gun (using air) out of the garage and the six sensors demonstrated that the ball got up to 500 mph (250 m/s) in 8 inches, then slowed down as it proceeded down the rest of the barrel.

We shattered a lot of ping-pong balls before the urethane idea occurred just 10 days before leaving for M.I.T. Then it still failed until we discovered it took a week to cure the plastic; so we never had balls that worked until the week of the demonstration. It was a good thing we put the plastic in a dozen balls at the outset.

THE WEEK

Monday

Pipe arrives. It rains. We assemble the 1000 foot pipe (saving a few lengths in case a truck should run over a section).

Tuesday

Pulled a vacuum;tried a shot. Ball got stuck in tube. Found stones and plastic pieces of the pipe in the tube. (Ball hitting stone at high rate really gets messed up!) Found vacuum cleaner; blew a string through and pulled a wad of cloth through to clean out pipe.

The wire from the far end back to the computer had so much capacitance the amplifier would not drive it so we gave up trying to measure final velocity. (It did not matter since there was hardly any final velocity anyway.) Could not get strobe lights to place down the pipe to indicate passage of the projectile to the waiting observers. (Saved having to rig up power supplies down the pipe, though.)

The pipe undulated back and forth and up and down like a crooked snake; hard to see how anything could get through the myriad of wiggles. Someone got the idea that we get wood chips delivered and use them for a bed for the pipe, which we did in the morning. Joe spent all day Wednesday patting and straightening with great diligence.

The other two shots of the day did not work either so we we decided to pump the vacuum all night. (It takes about an hour to pump the pipe down to the point that the gauge stops moving.)

Wednesday

First shot of the day was disappointing. The diaphragm broke at only 24 psi and the shot ended up in the middle, had to be blown out. We deduced that the diaphragm had stretched all night and was weak. The vacuum was still not very good--0.1 lb/in^2--we needed .01 or less.

Finally evacuated and made two shots. The first with 45 lb/in^2 and 0.005 acetate, the ball did not make it. The second shot, with a bit better vacuum (but still not good, 0.2 lb/in^2 indicated) got to within 20 feet of the end. Felt better. Decided to double the size of the breech so that there would be more energy given to the shot. (From then on we could not measure the speed because it exceeded what the photocells would measure.)

Thursday (the day of the demo)

The first shot did not get to the end either. Beginning to worry. Decided that there must be leaks but we could not figure out how to find them so we painted all the joints again with pipe glue.

The second shot did not make it, or so we thought because the strobe light at the end did not flash, but then a guy on a bicycle showed up who said "There is a ball in the tube at the end" and we cheered. This is the shot that was shown on TV (including the cheering). Now it was about 1 P.M. and the demonstration was supposed to take place at 4 P.M. A spectator said, "Why don't you pressurize the tube and paint it with soap solution and find the leaks." We said "Vacuum pumps will not pressurize well (too much oil, wrong fittings)." He said, "Just use your helium to do it!" Wow! Wish we had thought of that months earlier back in Albuquerque.

Some of the audience ran off to the store for Joy dishwashing detergent and paper cups and we all set to work painting and found 7 leaks. Had a nightmare: what if we broke the fittings while tightening them? Got things fixed up by 3:30 and it took an hour to pump down. Fortunately the bus was late.

Shot the shot, still no strobe . . . failure. But people went to take a look and the ball had made it, the strobe had failed . . . Success. But no panache whatsoever.

APPENDIX

First letter report to Macro-Engineering Research Group in September, 1984, before any work was done at all.

PRELIMINARY IDEAS FOR MAKING THE DEMONSTRATION TUNNEL

It will be hard to do for $6000, maybe impossible (eventually we did it for $1200). Standard parts must be used. 2-inch plastic pipe is available, clear, at $4 per foot and we can send a carrier such as banks use in their vacuum tellers down it at 1500 f.p.s. with a light on the front. Good at night anyway.

Cannot possibly accelerate it at nominal rates: 1 G would take 7 miles to get to 1600 km.p.h. (1000 m.p.h., 1500 f.p.s.). 70 G could do it in 500 feet. The magnetics people (John Williams) say this is fairly easy, so I will try for greater accelerations. (If so, things get shorter and cheaper). Then we just let it coast between the ends. Then we do it with the air in the tunnel and watch it slow down in a hurry. If it gets stuck, we blow it out.

I understand this has been done with 20-foot demonstrations but I agree that you need to show it over one or two thousand feet.

Surely magnetic levitation would be far too expensive. Let us use air lubrication. If there were two "wear rings" at each end separated by a small space with air introduced in the space, that should do it.

You would accelerate it magnetically however. Charge up some NiCad batteries (only a few seconds life needed) to make the carrier a magnet, put in some compressed air for the bearings, put the thing in the end of the tube and . . . fire! Stability of the air bearings would be a consideration that would play into the length of the carrier.

You would make sure that if it did not get stopped at the far end, it would just punch through a diaphragm.

So you end up with 1000-2000 feet of clear tube, with a 500 (or less) foot section ringed with coils on each end and a further section of copper tube perhaps needed to slow it down by eddy current drag before it came out the end. The coils would each have a capacitor maybe, charged between runs and also from the deceleration (one diode per coil), and the capacitor would be fired into its coil by a photocell watching for the carrier to approach, looking through the tube. By spacing the coils a bit people could see it accelerating (at night). If the copper tube were prohibitive in price, perhaps we could dynamically decelerate the thing with the propulsion coils at the receiving end. (We eventually did it much simpler with the gas gun, a suggestion of a physicist friend of mine in California.)

There is a problem making the tube super straight. How about floating it in a trough full of water (another tube, bigger and split in half) and stretching the two ends (a lot--plastic expands a lot in the heat so the length would change quite a bit anyway). The water takes care of the main support problem--gravity. The stretching helps the sideways bends (though not completely, the wear rings will indeed wear, I am sure).

I will be going to Rensselaer Polytechnic Institute next Monday and will talk to them; to Boston on Tuesday through Friday and will talk to M.I.T. folk; then I will talk it up here in the state. Our governor is interested in high-speed rail, maybe we can get a little more cash. But first I have to understand how to magnetically accelerate things. (Some R.P.I. professors felt the whole thing was a fraud on the public because when you, inevitably, had to let air in the tube to rescue people, it would heat up and fry them.)

SECOND LETTER REPORT OCTOBER 1984

I got 120 feet of 2-inch plastic pipe laid out on the ground. The first 30 feet and last 20 feet was clear ($4/ft); the rest was opaque ($.50/ft). I bought a vacuum pump (for evacuating refrigeration systems) and pumped the pipe down for 20 minutes. A diaphragm at the "head" end of two layers of aluminum foil and one at the "foot" consisting of a thin sheet of aluminum glued on to the flange with rubber cement sealed the tube. When the head-end diaphragm was poked, air in the atmosphere flooded the pipe with a bang and pieces of foil were blown to within 2 feet of the far end. Then we put in a balsa "bullet" 3 inches long, evacuated and fired again. The end diaphragm flew twenty feet and the projectile was not to be found!

The next week we purchased a mercury manometer so that we could measure the vacuum and a couple of valves and installed them on the weekend. Sure enough, we discovered a leak or two. This time we built a "gun" consisting of a 1-meter-long section of tubing ahead of the "head" diaphragm that confined about a liter of atmospheric air (in future experiments it will be pressurized up to around 150 lb/in^2 perhaps) and a device to puncture the diaphragm. This time the bullet was a sphere of styrofoam (from the local florist shop). We pumped the apparatus down to 30 millimeters and fired away. Unfortunately, there was a piece of something caught in the pipe (we had left it open during the week) and nothing much happened. Upon blowing through the pipe to clear it we discovered that the obstruction was pretty complete--may be a piece of the balsa bullet? (It was!)

This weekend I got out my paint spray pump and made a "reamer" with a bolt and a couple of 2-inch washers, but my low volume paint pump would not push it through the pipe even with no obstruction visible and I could only spend an hour or so on the project because, as noted, I got the flu.

The next step is to get a better reamer made up with pump leathers to seal it and really clean out the pipe. If necessary cut the pipe and reseal it to get the job done. Then try the styrofoam again. Leather did not work either. We cut open the tube and found the balsa bullet in the middle and pushed it out. The problem of having the projectile end up in the middle somewhere kept recurring.

Meanwhile, I have some theoretical studies going at N.M.S.U. and at N.M. Tech and at U.N.M (a little competition, you see) in an effort to figure out how to lubricate the bullet and to get the expansion of the gas from the gun figured out, so as to size the gun and get its pressure right. The lubrication people have suggested that I shoot an ice cube (cylindrical, hollow) that would melt and evaporate (below 6 millimeters pressure) and lubricate very well perhaps. Another idea is to use a ping pong ball for a projectile. The idea is to have something very light and well understood by the laymen and watch it go very far with a very little puff of air to start it. I talked to a couple of laymen who did say that they had experienced lots of bullets at high speed and would be much more impressed with something lightweight (and obviously very easy to slow down, in air) going very far.

Another suggestion that was put forth was to make the tubing into a giant circle (perhaps 100-meter diameter) and watch the projectile go around and around.

A little more of the gas dynamics: The speed of sound in hydrogen is much faster than in air; maybe in the end we should accelerate with that to get extreme velocities.

I will report again in two weeks, hopefully with a bit better results. I am saving the bills but I have about $450 in pipe and fittings, $220 in the vacuum pump, and $280 in the manometer plus about [illegible]50 in miscellaneous, so I have about used up your money at this point. You own (or I will buy back from you in the end) a manometer and vacuum pump. I expect to use up only as much more cash as it might take to get the speed measuring stuff (two places adjustable about 1/3 and 2/3 points) working (I can borrow some time measuring equipment, use a microcomputer --Rockwell Aim 65--that I own and maybe only get photocells and light-emitting diodes). Only when the speeds are right, the projectile is figured etc., and I get your approval will I spend more cash for tubing and at that time you get videos, movies, and/or still pictures of the set up as a minimum. (We did make a rather boring videotape.)

REPORT TWO WEEKS BEFORE THE SHOW

At this time (April 3, 1985), a ping-pong ball (38 millimeters outside diameter, weighing 3 grams) has been accelerated to 250 m/s using air as the propellant with the gun shooting into the atmosphere using a mylar diaphragm that broke about 420 kPa (62 psi). A similar shot using helium as a propellant realized over twice the muzzle velocity (546 m/s). The ball accelerated in about 8 inches showing about 15,000 G's acceleration.

Repeating the helium acceleration with the gun projecting into the vacuum tunnel yielded a shattered ping pong ball. The velocity could not be measured, of course. A search is on to find a light yet strong projectile. The ball shattered within a few inches of being hit by the shock by the mechanism of the rear overtaking the front of the sphere and passing through it. The calculated speed of the shock was 740 m/s. (We tried handballs, nurf balls, golf balls . . . everything was too heavy to accelerate and the ping-pong balls broke.)

Earlier experiments with balsa wood bullets, ice cubes, and styrofoam balls which were the same size as the bore of the tube and using air as the propellant resulted in no bullet getting to the far end and, due to the action of the (small amount of) air in the tunnel being compressed ahead of the bullet, stopping it and sending it back up the tube. Even though the first and last 10 meters of the tube are transparent and observers were present, no one saw any of the action except in the case of the very heavy ice cube which oscillated from one end of the 40 meter test tunnel to the other several times. (We solved the problem by using larger tube and smaller balls and using a smaller bore "gun" to shoot into the larger pipe. That way whatever air was ahead of the shot could get out of the way instead of being compressed. We also put a large diameter section of pipe at the far end as a "catcher" and put some dirty socks in it.)

REPORT JUST AFTER THE SHOW (APRIL 1985)

As usual, you think of what you should have done after it is all over . . .

I think the reason that we did not get a reading of speed on the last three shots (all the Thursday ones) was that the bullet was going too fast for the sensors. Since I had already clocked speeds over 500 m/s the implication is that we were going even faster. This would be because I increased the length (volume) of the breech the last thing Wednesday night.

It may be that we had a better vacuum than I thought. This would help explain why the ball got to the end even though the previous shot, which barely made it, read a lower pressure. I have a hunch that the pressure gauge was not an absolute pressure gauge after all but rather that it measured the differential pressure between the atmosphere and the inside of the pipe. The atmospheric pressure probably was changing over time. Also, we "calibrated" it rather crudely: we pulled a vacuum on it alone with the big pump and called that "zero" pressure . . . but even then, John said that he did not get it to read exactly zero. (We checked later and sure enough, it was not an absolute pressure gauge. We also decided that the atmospheric pressure had been going down all week. If it had gone up the gauge would have gone negative and we might have gotten a clue . . . then again, maybe we would not have.)

I think the reason that the strobe did not flash was that the tube down there was all scratched up and opaque. I should have set up the photocells with my meter as I did the ones at the beginning. Then I hould have put somebody at the end to just wave if it got there, also! Always have a plan "B.")

OTES FOR THE NEWS

I should have said that it was an effort to demonstrate low energy chemes to allow fast transport instead of saying that you first need a traight roadbed and second to start getting rid of the friction . . .

IST OF CREW THAT ASSISTED THOMAS STOCKEBRAND IN PREPARATION OF MODEL

Several wonderful people helped through all this, notably Joe Hodnick nd the technician John who was on duty 100% of the time, and the hotographer who had been sent up from New York for three days by *ime-Life*--real godsends.

The crew who assisted me in preparing and placing the model included:

Professor David G. Wilson, Department of Mechanical Engineering, Chief of Staff,

Sarah Jane Neustadtl, former Managing Editor, *Technology Review*,

Joseph Hodnick, M.I.T. civil engineering graduate, now an artist and specialist in vision analysis,

Suzanne Fairclough, former Associate Editor, *Technology Review*, now curator, Macro-Engineering Research Group Film Collection,

Paula Korn, freelance writer and former television director, *Financial News*, California; consultant to Macro-Engineering Research Group,

Robert Bennett, urban planner, M.S. from University of Massachusetts (Amherst),

Brian Barth, M.I.T. sophomore, who also assisted Professor Wilson with his model of the patented PAT system, exhibited at the Copley Plaza Hotel during the Conference

Index